T3-BSF-311

Monte Carlo Modeling for Electron Microscopy and Microanalysis

OXFORD SERIES IN OPTICAL AND IMAGING SCIENCES

1. D. M. Lubman (ed.). *Lasers and Mass Spectrometry*
2. D. Sarid. *Scanning Force Microscopy With Applications to Electric, Magnetic, and Atomic Forces*
3. A. B. Schvartsburg. *Non-linear Pulses in Integrated and Waveguide Optics*
4. C. J. Chen. *Introduction to Scanning Tunneling Microscopy*
5. D. Sarid. *Scanning Force Microscopy With Applications to Electric, Magnetic, and Atomic Forces, revised edition*
6. S. Mukamel. *Principles of Nonlinear Optical Spectroscopy*
7. A. Hasegawa and Y. Kodama. *Solitons in Optical Communications*
8. T. Tani. *Photographic Sensitivity: Theory and Mechanisms*
9. D. Joy. *Monte Carlo Modeling for Electron Microscopy and Microanalysis*

Monte Carlo Modeling for Electron Microscopy and Microanalysis

DAVID C. JOY

New York Oxford
OXFORD UNIVERSITY PRESS
1995

Oxford University Press

Oxford New York
Athens Auckland Bangkok Bombay
Calcutta Cape Town Dar es Salaam Delhi
Florence Hong Kong Istanbul Karachi
Kuala Lumpur Madras Madrid Melbourne
Mexico City Nairobi Paris Singapore
Taipei Tokyo Toronto

and associated companies in
Berlin Ibadan

Published by Oxford University Press, Inc.,
200 Madison Avenue, New York, New York 10016

Library of Congress Cataloging-in-Publication Data
Joy, David C., 1943—
Monte Carlo modeling for electron microscopy and microanalysis /
David C. Joy.
p. cm. — (Oxford series in optical and imaging sciences : 9)
Includes bibliographical references.
ISBN 0-19-508874-3
1. Electron microscopy—Computer simulation. 2. Electron probe microanalysis—Computer simulation. 3. Monte Carlo method.
I. Title. II. Series.
QH212.E4J67 1995
502′.8′25—dc20 94-35642

A disk, 3½-inch MS-DOS format, containing all the source code in this book is available by mail from David C. Joy, P.O. Box 23616, Knoxville, TN 37933-1616, U.S.A. The cost, including postage and handling, is $10.00 (U.S. and Canada) or $15.00 (elsewhere). Payments accepted in the form of a check or a money order.

1 3 5 7 9 8 6 4 2

Printed in the United States of America
on acid-free paper

PREFACE

Electron microscopy and electron beam microanalysis are techniques that are now in daily use in many scientific disciplines and technologies. Their importance derives from the fact that the information they generate comes from highly localized regions of the specimen, the data produced are unique in scope, and the images and spectra produced can be quantified to give detailed numerical data about the sample. For quantification to be possible, however, it is necessary to be able to describe how a beam of electrons interacts with a solid specimen—and such a description is difficult to provide because of the very varied and complex nature of the interactions between energetic electrons and solids.

The purpose of this book is to demonstrate how this interaction can be accurately simulated and studied on a personal computer, by applying simple physical principles and the mathematical technique of Monte Carlo (or random-number) sampling. The aim is to provide a practical rather than a theoretical guide to this technique, and the emphasis is therefore on how to program and subsequently use a Monte Carlo model. The bibliography lists other books that cover the mathematics of Monte Carlo sampling and the physical theory of electron scattering in detail. To make the programs developed here as accessible as possible, a disk—for use with MS-DOS–compatible computers—has been made available; it contains all of the source code described in this book together with executable (i.e., runnable) versions. To order see facing copyright page.

This book would not have been possible without the generous cooperation of many other people. I am especially grateful to Dr. Dale Newbury of N.I.S.T., for first introducing me to Monte Carlo models, and to him and Dr. Robert Myklebust, also of N.I.S.T., for sharing their code with me. Dr. Hugh Bishop of A.E.R.E. Harwell, whose Ph.D. work produced the first electron beam Monte Carlo programs, kindly lent me a copy of his thesis and provided some invaluable background information. Dr. Peter Duncumb, of the University of Cambridge and Tube Investments Ltd., whose pioneering work on Monte Carlo modeling using minicomputers ultimately made this book possible, lent me copies of his early reports and papers and has been unfailingly helpful in answering many questions about the development of the technique. The programs given in this volume have been refined and improved through the efforts of many colleagues who have used them over the past few years. Vital improvements in science, substance, and style have been made by

Drs. John Armstrong (California Institute of Technology), Ed Cole (Sandia), Zibigniew Czyzewski (University of Tennessee), Raynald Gauvin (Université de Sherbrooke), David Howitt (University of California), David Holt (Imperial College, London), Peggy Mochel (University of Illinois), John Russ (North Carolina State University), and Oliver Wells (IBM). To them, to my students Suichu Luo, Xinlei Wang, and Xiao Zhang, and to many others who have given of their time and expertise, I am deeply grateful. Any errors and problems that remain are strictly my own responsibility.

Finally, I dedicate this book to my wife Carolyn, without whose love and encouragement this manuscript would have remained just another pile of floppy discs.

Knoxville
August 1994

D. J.

CONTENTS

Monte Carlo Modeling for Electron Microscopy and Microanalysis

1
AN INTRODUCTION TO MONTE CARLO METHODS

1.1 Electron beam interaction—the problem

The interaction of an electron beam with a solid is complex. Within a distance of a few tens to a few hundreds of angstroms of entering the target, the electron will interact with the sample in some way. The interaction could be the result of the attraction between the negatively charged electron and the positively charged atomic nucleus (and equally the repulsion between the negatively charged atomic electrons and negative charge on the incident electron), in which case the electron will be deflected through some angle relative to its previous direction of travel, but its energy will remain essentially unaltered. This is called an *elastic* scattering event. Alternatively, the incident electron could cause the ionization of the atom by removing an inner-shell electron from its orbit, so producing a characteristic x-ray or an ejected Auger electron; it could have a collision with a valence electron to produce a secondary electron; it could interact with the crystal lattice of the solid to generate phonons; or, in one of several other possible ways, it could give up some of its energy to the solid. These types of interactions in which the electron changes both its direction of travel and its energy are examples of *inelastic* scattering events. After traveling a further distance, the electron will then again be scattered, either elastically or inelastically as before, and this process will continue until either the electron gives up all of its energy to the solid and comes to thermal equilibrium with it or until it manages to escape from the solid in some way.

While at a sufficiently atomistic level this train of scattering events is presumably quite deterministic—given sufficient information about the electron and the parameters describing it—to an observer able to watch the electron as it travels through the solid, the sequence of events making up the trajectory for any given electron would appear to be entirely random. Every electron would experience a different set of scattering events and every trajectory would be unique. Since, in a typical electron microscope, there are actually about 10^{10} electrons impinging on the sample each second, it is clear that there is not likely to be any simple or compact way to describe the spatial distribution of the innumerable interactions that can occur or the various radiations resulting from these events. At best it will be possible to assign probabilities to specific events, such as the chance of an electron being backscattered (i.e., being scattered through more than 90°) or of being transmitted through the target; but any more detailed analysis of the interaction will be impossible. The Monte Carlo method described here uses such probabilities, together with the idea of sampling by using random numbers, to compute one possible set of scattering events for an electron as it travels into the solid. By repeating this process many times, a statistically valid and detailed picture of the interaction process can be constructed.

1.2 The Monte Carlo method

One of the very earliest published papers on "Monte Carlo" methods (Kahn, 1950) provides an excellent statement of the basis of the method—"By applying random sampling techniques to the problem [of interest] deductions about the behavior of a large number of [electrons] are made from the study of comparatively few. The technique is quite analogous to public opinion polling of a small sample to obtain information concerning the population of the entire country." The use of random sampling to solve a mathematical problem can be characterized as follows. A game of chance is played in which the probability of success P is a number whose value is desired. If the game is played N times with r wins then r/N is an estimate of P. The "game of chance" will be a direct analogy, or a simplified version, of the physical problem to be solved. To play the game of chance on a computer, the roulette wheel or dice are replaced by random numbers. The implication of a "random" number is that any number within a specified range (usually 0 to 1) has an equal probability of being selected, and all the digits that make up the number have an equal probability (i.e., 1 in 10) of occurring. Thus to take a simple and relevant example, consider an electron that can be scattered elastically or inelastically, the probability of either occurrence being determined by its total cross section. If the probability that a given scattering event is elastic is p_e and p_i is the probability of an inelastic scattering event (and the sum of p_e and p_i is unity), then a choice could be made between the two alternatives by picking a random number RND ($0 < \text{RND} < 1$) and specifying that if RND $\leq p_e$, then an elastic event occurs, otherwise an inelastic event is assumed to have occurred. If this selection procedure is applied a large number of times, then the predicted ratio of elastic to inelastic events will match the expected probability derived from the given probabilities p_e and p_i, since a fraction p_e of the random numbers will be $\leq p_e$. Random numbers can also be used to make other decisions. For example, if the probability $p(\theta)$ of the electron being scattered through some angle θ is known, either experimentally or from some theoretical model, then a specific scattering angle α can be obtained or picking another random number and solving for α the equation:

$$\text{RND} = \frac{\int_0^{\alpha} p(\theta)\, d\theta}{\int_0^{\pi} p(\theta)\, d\theta} \tag{1.1}$$

which equates RND to the probability of reaching the angle α given the known distribution of $p(\theta)$. By repeating these random-number sampling processes each time a decision must be made, a Monte Carlo simulation of one particular electron trajectory through the solid can be produced. The result of such a procedure is not necessarily or even probably a trajectory representing one that could be observed experimentally under equivalent conditions. However, by simulating a sufficiently large number of such trajectories, a statistically significant mixture of all possible scattering events will have been sampled and the *composite result* will be a sensible approximation to experimental reality.

1.3 A brief history of Monte Carlo modeling

The first published example of the use of random numbers to solve a problem is probably that of Buffon, who—in his 1777 volume *Essai d'Arithmetique Morale*—described an experiment in which needles of equal length were thrown at random over a sheet marked with parallel lines. By counting the number of intersections between lines and needles, Buffon was able to derive a value for π. Subsequently, other mathematicians and statisticians followed Buffon's lead and made use of random numbers as a way of testing theories and results. Because many of the phenomena of interest to physicists in the early twentieth century, such as radioactive decay or the transmission of cosmic rays through barriers, displayed an apparently random behavior, it was also an obvious step to try to use random numbers to investigate such problems. The procedure was to model, for example, a cosmic ray interaction by permitting the "particle" to play a game of chance, the rules of the game being such that the actual deterministic and random features of the physical process were exactly imitated, step by step, by the game and in which random numbers determined the "moves."

During the Manhattan Project, which led to the development of the first atomic bomb, John von Neumann, Stanislav Ulam, and others made innovative use of both random-number sampling and game-playing situations involving random numbers as a way of studying physical processes as diverse as particle diffusion and the probability of a missile striking a flying aircraft. It was during this period that these techniques were first dubbed "Monte Carlo methods" (Metropolis and Ulam, 1949; McCracken, 1955). Because the Monte Carlo method needs a large supply of random numbers as well as much repetitious mathematical computation, the later development of the technique was closely geared to the development of "automatic computing machines." As first mechanical and then electronic machines became available during the 1950s, the technique found increasing application to problems ranging in scope from diffusion studies in nuclear physics to the modeling of population growth by the Bureau of the Census. A valuable bibliography of these early papers and techniques can be found in Meyer (1956).

Although Monte Carlo methods had been applied to many other phenomena, it was not until the work of Hebbard and Wilson (1955) that the method was successfully employed for charged particles. Later work by Sidei et al. (1957) and Leiss et al. (1957) led to a major paper by Berger (1963) that laid the groundwork for future developments. Simultaneously, in England, M. Green, a physics graduate student in Cambridge working for V. E. Cosslett, was persuaded to investigate the application of von Neumann's Monte Carlo method, and the university's EDSAC II computer, to the scattering of electron beams. Taking experimental data on the scattering of electrons in a 1000-Å film as a starting point, Green (1963) and later Bishop (1965) were able to derive the electron backscattering coefficient, and the depth dependence of characteristic x-ray production, from a bulk sample and to demonstrate good agreement with measured values. However, while this approach showed the validity of the technique, it was limited in its application to those situations where the suitably detailed initial experimental data were available. At the 4th International Conference on X-ray Optics and Microanalysis in Paris in 1965, however, two independent papers (Bishop, 1966; Shimizu et al., 1966) demonstrated how theoretically based electron scattering distributions could replace and so generalize experimental distributions; within a short time, groups in Europe, Japan, and the United States had produced working programs based on this concept. One of the most important of these was the one produced by Curgenven and Duncumb (1971) working at the Tube Investments Laboratory in England. This program introduced several new concepts, including the so-called multiple scattering approximation discussed in Chapter 4 of this book, and was optimized to run on a relatively small scientific computer. Copies of the FORTRAN code were generously made available to interested laboratories throughout the world for their own use; as a result, this program came into widespread use and made a significant contribution to popularizing the idea that electron-solid interactions could be modeled conveniently and accurately by computer.

By 1976, the use of this technique was sufficiently common for a conference entitled "Use of Monte Carlo Calculations in Electron Probe Microanalysis and Scanning Electron Microscopy" to be held at the National Bureau of Standards (NBS) in Washington, D.C. The proceedings of that meeting (Heinrich et al., 1976) still form one of the basic resources for information in this field, and programs, algorithms, and procedures developed by the NBS group have formed the starting point for many of the programs in current use, including those described in this volume.

Apart from the very earliest examples, which were run by hand, using random numbers generated by spinning a "wheel of chance" (Wilson, 1952) or on mechanical desk calculators (e.g., Hayward and Hubbell, 1954), Monte Carlo programs were designed to be run on the main-frame computers then becoming available in most government and industrial laboratories and universities. Although such machines were both large and expensive, their capabilities were very limited, and considerable ingenuity was necessary to produce workable programs within the limits set by the available memory (often as little as 2000 words) and operating time between crashes of the system. Nevertheless Monte Carlo programs were often cited as a prime example of the new analytical power made available through electronic computing. With the advent of personal computers (PCs), this power is now available to anyone who needs it. Monte Carlo programs are, in most cases, relatively short in length and can readily be run on any modern PC without encountering any problems with the lack of memory. The programs are also computationally intensive, in the sense that once the program has obtained all the necessary data, it performs calculations continuously until its task is finished. This is not the ideal situation for a program that is run on a time-shared main-frame computer because it means that the actual computing time will depend directly on the number of users working on the machine at any given time (unlike programs such as word processing, where the majority of the computer's time is spent in waiting for the operator to enter the next character and multiple users produce little apparent drop in response speed). Consequently, even on relatively powerful time-share systems, the computational speed experienced by a user when running a Monte Carlo program can seem very slow; thus to calculate a sufficient number of trajectories to produce an accurate result can cost a lot of both time and money. While PCs do not, in general, perform the individual computations as rapidly as the main frame, they *are* dedicated to one task; as a result, their effective throughput can easily rival that of much larger machines. Also, since access to PCs is often free or at least very cheap, over-lunch or even overnight runs are no financial burden to the user. Finally, the interactive nature of PCs and the ready access to graphical presentation that they provide offers the chance to make programs that are both more accessible and more immediately useful.

1.4 About this book

This volume is not intended to replace standard textbooks on the general theory of Monte Carlo sampling (such as those by Hammersley, 1964, and Schreider, 1966), nor is it a substitute for a comprehensive guide to electron beam microscopy and microanalysis (such as Goldstein et al., 1992). Rather, it is intended to provide electron microscopists, microanalysts, and anyone concerned with the behavior of electrons in solids with ready access to the power of the Monte Carlo method. It therefore provides working Monte Carlo simulation routines for the modeling of electron trajectories in a solid and discusses procedures to deal with associated phenomena such as secondary electron and x-ray production. These procedures can then be added to the basic simulation as required to produce a program customized to tackle particular problems in image interpretation or microanalysis. The goal is to make available simulations whose accuracy is at least as good as that likely to be achieved in a comparable measurement or experiment on an electron microscope. The programs have been developed and have been designed to be run on personal computers rather than on scientific minicomputers or full size main-frame machines. Even given the advanced PC designs now available, this has occasionally made it necessary to compromise between the completeness of the model and the speed of execution. In most cases the choice has been for the version that is fast in operation, since a good approximation available rapidly is much more useful than an exact result that takes a day to compute. No claim is made that these programs represent the best or even the only way to do the job. Indeed, a large number of other approaches are cited in the text. This book will have achieved its purpose if you—the user—feel ready, willing, and able to use the printed programs given here, or those available on the accompanying disk, as the basis for your own experimentation and development.

2
CONSTRUCTING A SIMULATION

2.1 Introduction

In this chapter, we will develop a Monte Carlo simulation of a random walk (sometimes called a "drunken walk," after its most popular mode of experimental investigation). Although this particular problem is only loosely related to the studies of electron beam interactions that follow, the model that we will develop provides a convenient way of establishing a framework for those subsequent simulations. It also illustrates the general principles of programming to be followed in this book and introduces some of the important practical details associated with constructing and running such models on a personal computer.

2.2 Describing the problem

The random walk problem can be stated as follows: "How far from the starting point would a walker be after taking N steps of equal length but in randomly chosen directions?" In order to simulate this problem, we must break it down to a sequence of instructions, or algorithms, that allow us to describe it mathematically. Figure 2.1 shows the situation for one of the steps making up the walk. It commences from the coordinate (x, y) reached at the conclusion of the previous step and is made at some random angle A with the X-axis, so that:

$$A = 2\pi * \mathrm{RND} \tag{2.1}$$

where RND is a random number between 0 and 1. The coordinates xn, yn of the end of the step are then:

$$xn = x + \mathrm{step} * \cos(A) \tag{2.2}$$

$$yn = y + \mathrm{step} * \sin(A) \tag{2.3}$$

Equations (2.2) and (2.3) can be cast in a more symmetrical form by writing

$$B = (\pi/2) - A \tag{2.4}$$

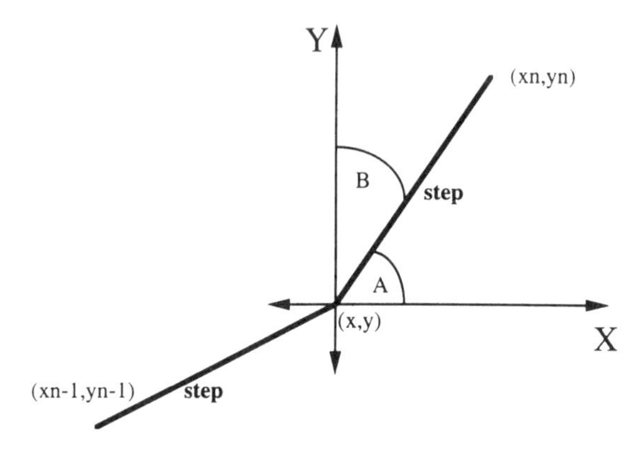

Figure 2.1. Coordinate system for the random walk simulation.

so that

$$xn = x + \text{step}*\cos(A) \tag{2.5}$$

$$yn = y + \text{step}*\cos(B) \tag{2.6}$$

cos (A) and cos (B), the cosines of the angles between the vector representing the direction of the step and the axes, are called the direction cosines and will in future be abbreviated to CA and CB.

Although Eqs. (2.1), (2.5), and (2.6) give an accurate mathematical description of one step of the random walk, the axes of the coordinate system are constantly changing as the walker moves from one step to the next. It would intuitively be more satisfactory to describe the progress of the walker with respect to a fixed reference frame of axes (such as the walls of the room), because this makes it possible to predict when, for example, a collision might occur. With this description then, as shown in Fig. 2.2:

$$\theta = 2\pi*\text{RND} \tag{2.7}$$

$$A = X + \theta \tag{2.8}$$

$$B = (\pi/2) - A = Y - \theta \tag{2.9}$$

where X, Y are the angles described by the direction cosines CX, CY for the previous step, and A, B are the angles for the new direction cosines CA, CB. As before

$$xn = x + \text{step}*CA \tag{2.10}$$

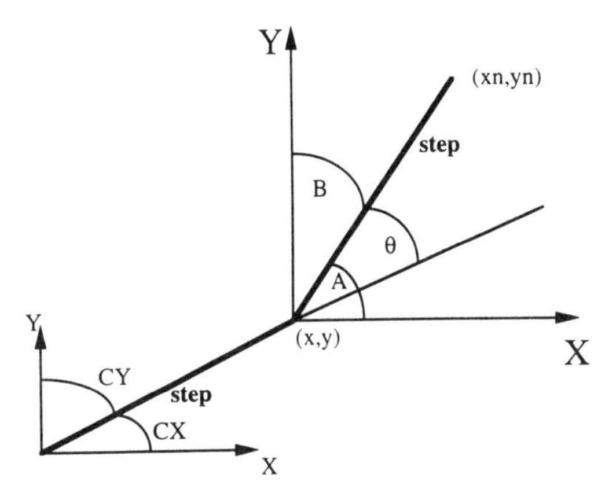

Figure 2.2. Modified coordinate description using fixed axes for the random walk.

$$yn = y + \text{step}*CB \tag{2.11}$$

and from the usual trigonometric expansions we get

$$CA = \cos(X + \theta) = CX*\cos(\theta) - CY*\sin(\theta) \tag{2.12}$$

$$CB = \cos(Y - \theta) = CY*\cos(\theta) + CY*\sin(\theta) \tag{2.13}$$

using the result that sin (X) = cos (Y) and sin (Y) = cos (X). Equations (2.7), (2.10), (2.11), (2.12), and (2.13) now describe how to calculate the end point of a step, given its starting point. The next step can similarly be computed using the identical equations but resetting the coordinates so that the old end point becomes the new starting point and the exit direction cosines become the entry direction cosines:

$$x = xn,\ y = yn,\ CX = CA,\ CY = CB \tag{2.14}$$

The recipe for simulating the random walk is therefore as follows:

Given a starting point (x_0, y_0) and a starting direction (CX, CY), then
Repeat the sequence
Find the deviation angle θ [Eq. (2.7)]
Calculate the new direction cosines CA, CB [Eqs. (2.12) and (2.13)]
Calculate the new coordinates xn, yn [Eqs. (2.10) and (2.11)]
Then reset the coordinates for the next step
$x = xn,\ y = yn,\ CX = CA,\ CY = CB$
until the required number of steps has been taken
Then distance s from starting point is $s = \sqrt{(x - x_0)^2 + (y - y_0)^2}$

2.3 Programming the simulation

To carry out the simulation on a computer, the recipe given above must be expressed in a form that the computer can understand. This requires that we choose one computer language, from among the many now available, in which to code our program. Unfortunately, any discussion of programming languages is liable to lead to acrimony, because anyone who regularly uses a particular language and has become used to its syntax and particular strengths and weaknesses can always find sufficient reasons to prove that any other language is deficient in power or convenience. However, stripped of the theological overtones so often accompanying this sort of debate, the truth is that any of the languages now in common use on personal computers could be used to code this and the following, Monte Carlo simulations without a noticeable effect on the quality of the final product. But since it is not practical to provide equivalent code for all possible languages, it is necessary to choose just one arbitrarily, for whatever reasons seem appropriate, and work with that.

Even though it might not have been your first or even your second choice, the decision here has been to use PASCAL. The reasons for this decision were principally as follows:

1. PASCAL is a good example of a modern language. It allows for structured and modular programming; it has a powerful yet simple syntax; and—since it is a compiled language—it is fast in execution.

2. The style of a PASCAL program, in particular the use of indenting and the availability of long descriptive variable names, leads to code that is easy to read and understand.

3. Variants of PASCAL are commercially available for all computers likely to be encountered in current use. Although there are slight differences between them, the original definition of the language was sufficiently precise that these variations rarely pose a problem if a program must be moved from one version of PASCAL to another.

4. Finally, if you cannot take PASCAL at any price, then—since other modern languages such as QUICKBASIC, ADA, MODULA-2, or C now share so many of the features of PASCAL—conversion from PASCAL to any other language of your choice is straightforward. In fact, software is available that can effect such transformations automatically in many cases.

The programs in this book are written in TURBO PASCAL™ (version 5.0), which is perhaps the most widely used form of PASCAL for MS-DOS computers. These programs will also compile and run, without any changes being necessary, in

Microsoft QUICK PASCAL™ version 1.0 and higher. Other variants of PASCAL and other types of computers may require some modifications to the code, especially for the graphics commands. A disk (IBM/MS-DOS format) containing all of the code discussed in this text, as both source code and as compiled and executable code, is available from the author. Ordering details are given on the copyright page of this book.

2.4 Reading a PASCAL program

The PASCAL program that implements the mathematical description derived above for the random walk simulation is as follows:

```
Program Random_walk;
   {this stimulates a simple random walk with equal-length steps}

uses CRT,DOS,GRAPH; {resources required}

var
   CA,CB,CX,CY:extended;                    {direction cosines}
   x,xn,y,yn,theta,distance:extended;       {step variables}
   step:extended;                           {display variables}
   hstart,vstart:integer;                   {screen center}
   i,tries:integer;                         {counter variables}
   GraphDriver:Integer;                     {which graphics card?}
   GRAPHMODE:Integer;                       {which display mode?}
   ErrorCode:Integer;                       {is there a problem?}
   Xasp,Yasp:word;                          {aspect ratio of screen}

cons
      twopi=6.28318;                        {2π constant}

Procedure set_up_screen;
   {gets the required input data to run the simulation}

  begin                                     {ensures screen is clean}
     ClrScr;
      GoToXY(10,5);
      writeln('Random Walk Simulation');

      GoToXY(33,5);                         {get number of steps}
         write('. . . . . . . . . how many steps?');
            readln(tries);
  end;

Procedure initialize;
   {identify which graphics card is in use and initialize it}
```

```
var
   InGraphicsMode:Boolean;
   PathToDriver:String;

  begin
   DirectVideo:=False;
    PathToDriver:='';
     GraphDriver:=detect;
    InitGraph(GraphDriver,GRAPHMODE,PathToDriver);
   SetViewPort(0,0,GetMaxX,GetMaxY,True);    {clip view port}

   hstart:=trunc(GetMaxX/2);     {horizontal midpoint of screen}
   vstart:=trunc(GetMaxY/2);     {vertical midpoint of screen}
   step:=(GetMaxX/50);           {a suitable increment}

   GetAspectRatio(Xasp,Yasp);    {find aspect ratio of this display}

  end;

Procedure initialize_coordinates;
   {set up the starting values of all the parameters}

  begin
               x:=0;
               y:=0;
              CX:=1.0;
              CY:=0.0;

              randomize; {and reset the random-number generator}
  end;

Procedure new_coord;
   {computes the new coordinates xn,yn given x,y, CX,CY and theta}

var
   V1,V2:extended;

  begin

   V1:=cos(theta);
   V2:=sin(theta);

     CA:=CX*V1 - CY*V2;    {new direction cosines}
     CB:=CY*V1 + CX*V2;

   xn:= x + step*CA;
   yn:= y + step*CB;       {new coordinates}

  end;
```

```
Procedure plot_xy(a,b,c,d:extended);

   {plots the step on the screen. Since all screen coordinates
    are integers, this conversion is made first. The real X,Y
    coordinates are separated from the plotting coordinates,
    which put X=Y=0 at the point hstart,vstart}

var
    ih,iv,ihn,ivn:integer;    {plotting variables}

  begin

    ih:=hstart + trunc(x*Yasp/Xasp);    {correct for aspect ratio}
    iv:=vstart + trunc(y);

    ihn:=hstart + trunc(xn*Yasp/Xasp);    {ditto}
    ivn:=vstart + trunc(yn);

    line(ih,iv,ihn,ivn);          {and draw the line}

  end;

Procedure reset_coordinates;
   {shifts coordinate reference X,Y to new coordinates XN,YN and
     resets the direction cosines}

  begin
               x:=xn;
               y:=yn;
              CX:=CA;
              CY:=CB;

  end;

Procedure how_far;
   {computes the distance traveled in the walk}
label hang;

var
    a,b:integer;
    s:string;

  begin

     distance:=sqrt((x-hstart)*(x-hstart) + (y-vstart)*(y-vstart));
      distance:=distance/step;

                              {in step-length units}
```

```
            a:=trunc(GetMaxX*0.2);    {adjust for your screen}
            b:=trunc(GetMaxY*0.9);    {ditto}

          Str(distance:3:1,s);    {convert number to text string}

            s:=concat(s, 'steps from origin');

    OutTextXY(a,b,concat ('The walker traveled' s));

   hang:                                      {label for goto call}

     if (not keypressed) then goto hang;     {keep display on screen}
          closegraph;                         {turn off graphics}

  end;

{*********************************************
 *            main program starts here          *
 *********************************************}

begin

   set_up_screen;

   initialize;    {find graphics card and set it up}

   initialize_coordinates;    {set up variables}

{*********************************************
 *              loop starts here                *
 *********************************************}

for i:=1 to tries do

  begin

      theta:=twopi*RANDOM;    {get deviation angle}

       new_coord;              {compute new position}

       plot_xy(x,y,xn,yn);     {plot this step on screen}

       reset_coordinates;      {move origin to XN,YN position}

       delay(100);             {to slow down display}
```

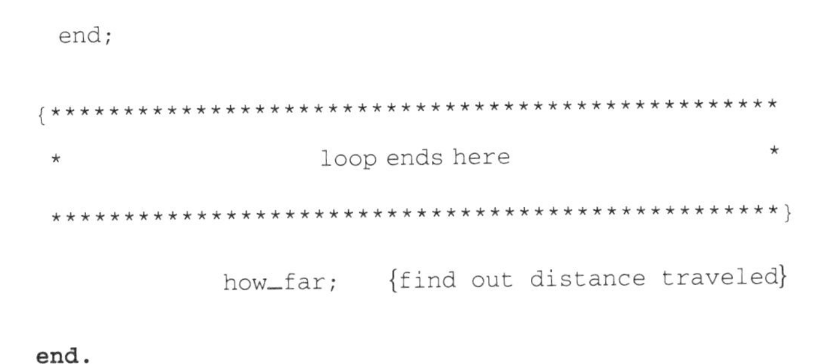

```
  end;

{*************************************************
 *                  loop ends here              *
 *************************************************}

           how_far;    {find out distance traveled}

end.
```

All PASCAL programs have exactly the same layout, which aids in following the code even though the format itself seems, at first, rather unusual. To further help the reader to follow the structure and logic of the program, this book will also employ a consistent set of typeface conventions when listing the programs in the text. If you decide to type any of these programs into your computer to run them, then these bold and italic effects should, of course, be ignored. The program will be set in `Courier typeface;` quotes from the program code, procedure, and function names, etc., will also be set in the same typeface to identify them in the text.

The first line in the listing always identifies the program name, here `RANDOM WALK`. The keyword `PROGRAM` will be printed in bold type to identify it. Note that this statement, like all PASCAL statements, ends in a semicolon (;). This is the "delimiter" that separates one program statement from the next. The PASCAL compiler ignores spaces and line breaks, following the program logic only from the delimiters. If desired, further details about the program can then be added as a comment. In PASCAL, this is done by enclosing the text within {. . .} brackets. For clarity in the printed listings, comments will be shown in italics.

Next, the resources required by the program are listed. The keyword **`uses,`** here printed bold, is followed by a list of the libraries required by the code that follows. In this case the CRT, DOS, and GRAPH libraries are needed. On other computers or in other variants of PASCAL, the list of resources may be different or missing altogether.

The keyword **`var,`** which will be in bold, introduces a listing of all the variables used in the program, grouped by their type. Every variable in PASCAL must be defined before it is used and be of a type (e.g., integer, longinteger, extended, real, Boolean, etc.) recognized by the program. Once a variable is "typed," its properties are fixed; so if for some reason a particular number is needed in more than one form (for example, as both a real number and an integer), two variables must be defined and the value converted from one type to the other, using one of the special functions provided for this by the language. Variables listed at the start of a program are GLOBAL variables; that is, every part of the program can read and write these quantities. Variables can also be defined within the procedures and functions used by the program (see below). In this case, the variables are local, or private, to the particular procedure or function in which they appear and are not available to other parts of the program. Note that PASCAL does not distinguish the case of letters, so dummy, DUMMY, and Dummy would all refer to the same quantity.

CONST identifies a list of parameters that will be of constant value in the program, such as the length of the step taken by the walker and the coordinates of the center of the display screen. Once defined here, these quantities are also global in their effect. Constants appearing in the **CONST** list do not appear in the **VAR** list, and their type is determined by the format assigned to them when they first appear. Thus

```
CONST        dummy = 5;
```

will define a constant of integer type, while the statement

```
CONST        dummy = 5.0;
```

would define one with properties of a real number. Once assigned, the value of a constant cannot be changed; any attempt to do so will lead the compiler to produce an error statement. Temporary constants—that is, ones whose values may be altered if required—are obtained by a normal program line equating two variables:

```
new_dummy:=5.15;
```

Note that assigning the value on the left-hand side of a statement to value given on the right-hand side requires a ":=." The simple "=" implies a logical operation, as in "if A=B," not the act of equating two values. Since it has been estimated that 95% of all errors in PASCAL come from omission of the ";" and confusion between "=" and ":=," care in copying these jots and tittle will be repaid by a reduction in the time it takes to get a program running.

Following these definitions, the listing contains code for **FUNCTION** and **PROCEDURE** routines. These pieces of code play the role assigned to subroutines in some other languages. A function has the specific job of performing some defined mathematical operation. It therefore returns to the program line that called it an output of some type, and this type must appear in the definition. Thus

```
Function SUM: real;
```

tells us that this is a function whose output is a real number—for example, the sum of two or more other numbers. If the function is supplied with data on which to work, then both the type of the data supplied and of the output returned must be given, so

```
FUNCTION black_box(a,b:integer):real;
```

takes in data defined as integer but produces as output a real number. Many functions, such as sin or cosin or log, are already included in the language itself; but by writing the necessary code, as many new and specialized functions as required can be added. A procedure, by comparison, can perform any legal set of operations and so has no "type" of its own. In either case, the procedures and functions must be defined before they are called by the program, so all of the code associated with these items appears before the main body of the program. As far as possible, all of the detail in a program is carried out by procedures and function calls. The main program then simply lists the order in which these are carried out. This makes the logic of the program easy to follow, particularly if descriptive names are used for the functions and procedures. Since PASCAL imposes no limitations on the length of names, it is worth choosing ones whose meaning will still be evident 6 months after a program is written (i.e., set_to_zero is descriptive and helpful, STZ is not). Laying out programs in this way also makes it easy to subsequently construct new programs, since the components can be reused.

The main program runs between the key words **BEGIN** and **END,** set in bold type. For clarity, the start will also usually be identified by a suitable line of comment. In reading a program for the first time, therefore, the main program code should be identified, remembering that it is always at the end of the listing; then the procedures and functions should be studied as they are called by the program. The program starts by calling the procedure `set_up_screen`. which, as can be seen from the code for it, first clears the display screen, presents the program name ready for the simulation to begin, and asks for the number of steps to be modeled.

Since the purpose of this program is to plot the random walk on the screen, the next step is to set up the graphics display of the computer. This is complicated by the fact that MS-DOS machines come with a wide variety of graphical display hardware. The procedure `initialize` (taken directly from the TURBO PASCAL reference books supplied by Borland Inc.) determines automatically which type of graphics display (i.e., CGA, EGA, VGA, Hercules, etc.) is present in the computer at the time the program is run. It then selects a "graphics driver program" (files on the disk with the extension .BGI) that will interface the computer code to the hardware in use. Provided that our software is written carefully enough, the same code will produce an acceptable output on any of the possible displays, even though these will differ in parameters such as their resolution and the shape of the screen.

The starting point of the random walk is to be placed at the center of the screen, but the coordinates of this point will vary from one type of display to another. We therefore ask the machine to tell us how many picture points ("pixels") it can plot in the X (or horizontal) direction, using the call `GetMaxX`, and how many in the Y (or vertical) direction, using `GetMaxY`. The center of the screen is then, quite generally, at the coordinates (GetMaxX/2, GetMaxY/2). In plotting on the screen, the coordinates must be given as integer number, so the procedure converts the real numbers (GetMaxX/2, GetMaxY/2) to integers (hstart, vstart) using the `'trunc'` function defined in PASCAL. Similarly, the step length is set to be one-fiftieth of the width of the screen in the horizontal direction (i.e., GetMaxX/50). A final touch is to find the "aspect ratio" of the display—that is, the ratio of a unit step in the horizontal direction to a unit step in the vertical direction. This is obtained from the call `GetAspectRatio()`, which gives us the ratio Xasp/Yasp for the display in use. By correcting for the aspect ratio of the screen, we can ensure that the displays on the screen will look similar for all the graphics cards supported by the software.

The next procedure used, `initialize_coordinates`, sets the position variables x,y and the direction cosines CX,CY to the desired starting values. In some computer languages (e.g., BASIC), all variables are automatically reset to zero each time the program is run; but in PASCAL, the value assigned to a variable when the program starts is quite arbitrary, so it is necessary to explicitly set each to the value required. The procedure also resets the random number generator with the statement "`randomize`." The quality of a Monte Carlo simulation depends to a great extent on the quality of the random numbers with which it is supplied, since if these exhibit any patterns in their behavior or if a given number repeats within a limited number of calls, the output data will be flawed. The random-number generators in TURBO PASCAL (and QUICK PASCAL), summoned by the call `RANDOM` are quite well behaved provided that the generator is periodically rerandomized using the call `RANDOMIZE`. Although it is not a documented feature, it can also be noted that if the initial `RANDOMIZE` statement is omitted (or placed within {. .} comment brackets), then the same string of random numbers will be generated every time the program is run. This can, in some cases, be useful for debugging a simulation and also in examining the effect of changing a single parameter in a simulation.

The program now calls the procedure `new_coord`, which calculates the new coordinates xn, yn given the starting coordinates x, y; the initial direction cosines CX, CY; the deflection angle θ; and the step length, using the formulas derived above. Because all of these variables have been declared to be global by being listed at the start of the program, no variables need be passed to the procedure. The effect of this step is then displayed on the screen, using the procedure `plot_xy`, which draws the step on the screen. Note that since the graphics routines that place pixels on the screen need integer numbers as input, the procedure must first convert the actual real number coordinates x, y, xn, yn to local integer variables ix, iy, ixn, iyn using the `trunc` function supplied in TURBO PASCAL. The origin of the simulation ($x = 0$, $y = 0$) is plotted at the center of the screen (hstart,vstart) and the coordinate system is such that a positive x value moves the point to the right and a positive y

value moves the point downward. As discussed in the procedure `Initialize`, the distance plotted in the *x* direction (horizontal) is corrected for the aspect ratio of the display screen in use, so that all screens will give roughly equivalent displays. Note that if the line being drawn has coordinates that would place it off the screen (that is if *ih* or *ihn* does not lie between 0 and GetMaxX, or if *iv* or *ivn* does not lie between 0 and GetMaxY), then the command `SetViewPort`, which appears in the procedure `initialize`, will automatically "clip" the plot at the edges of the CRT. The program will continue to run, but plotting will not resume until the line again lies in the visible region of the screen. The `plot_xy` procedure is one of the building blocks that we will be using again in later chapters. Next, the program calls the procedure `reset_coordinates`, which shifts the origin of the calculation from *x* to *xn*, and *y* to *yn*, and resets the direction cosines. A standard TURBO PASCAL command `Delay()` is then called. The purpose of this is simply to slow down the action of the loop so that it can be watched more easily on the screen. The delay time is approximately equal to the parameter passed to the function in milliseconds, thus `Delay(100)` will hold up the loop for about one-tenth of a second. The loop is then repeated until the desired number of steps has been calculated.

The program then exits from the loop, and the distance between the start and finish coordinates is computed using the procedure `how_far`. This uses Pythagoras's theorem to find the distance of the walker from the starting point in units a step in length. Since the computer is now in graphics mode, these data cannot be simply printed to the screen but must be drawn on the CRT as if it were a piece of a graph. First, the numerical value is converted to a string (i.e., a list of letters, digits, or symbols) using the standard TURBO PASCAL command `Str()`. The ":3:1" which follows the distance variable formats the result as three digits, one of which is after the decimal point. The string is assigned to a variable *s*, which is declared at the start of the procedure; it is therefore a local variable (existing only within this procedure). The string *s* is then concatenated (i.e., added) to other strings of text to form the message drawn on the screen by the command `OutTextXY()`. The message is drawn starting at the coordinates *a,b*, which are, again, local to this procedure. Since we wish one piece of code to serve for all types of display, the values of *a* and *b* are chosen as fractions of GetMaxX and GetMaxY. Since personal preferences about the appearance of the display may vary, however, the actual values can readily be changed to place the text at any other desired position on the screen. In fact, changing the values of *a* and *b* and then recompiling and rerunning the program to see the result is an excellent way for users without much practical experience with computers to gain proficiency in modifying the code and so to overcome any fear of the dire results of "interfering" with the program. Note that if *a* and/or *b* are chosen to be larger than GetMaxX or GetMaxY, the text may disappear because the command SetViewPort, in the procedure `initialize`, "clips" the display at the edges of the screen. In order to hold the display on the screen so that it can be examined or printed out, this procedure finishes with a loop.

The command `keypressed` calls a standard function that checks whether or not any key has been touched. The output of the function is a Boolean (i.e., it has the value true or false). In PASCAL, the code statement

```
if (function) then (operation)
```

will lead to the operation being performed if the function evaluates to a positive number or produces a Boolean variable with the value "true" (i.e., +1). If, on the other hand, the function evaluates to zero or to a negative number or produces a Boolean variable with the value "false" (i.e., 0) then the operation is not performed. Here, if keypressed is "false" (i.e., no key was touched) then the expression `not keypressed` has the value "true," so the program goes back to the label hang and continues to cycle around until eventually some key on the keyboard is struck and the procedure is exited.

It may seem to be overelaborate to present a simple program in this structured way, but adherence to this format produces code that is easy to follow. As the programs in this book become longer and more complex, the benefits of its use will be more evident. Debugging the program is also made easier because a whole procedure or function can be dropped out by simply placing {. . .} brackets around the call to it, allowing the errant region of the program to be quickly identified. In addition, by doing all of the real work in a program in the procedures and functions, a library of program modules is built up that makes it possible to put together a special-purpose program with the minimum of effort. Once a procedure or function has been debugged and tested in one program, it can safely be moved into another program with the certain knowledge that it will work as expected.

Finally it must be pointed out that the code, as given here, does violate one of the accepted principles of "good programming" in that global variables (i.e., variables declared at the beginning of the program and so visible to all of the procedures and function) are used. "Good" practice tries to avoid the possible side effects of global variables, such as the inadvertent corruption of a variable by a faulty procedure, by only passing copies of the variables to the procedures as they are required. However, in the electron beam simulations that follow, as many as 15 or 20 different variables may be required to describe the current status of the electron, and all of these may be needed by any given procedure or function. Writing the code to pass this number of variables to a procedure is cumbersome and the resulting operation is also slow, so the code given here respects the spirit of the injunction against global variables but uses them anyway in the interests of convenience and speed. For a more detailed discussion of this topic, for other questions related to programming in PASCAL, or for information on compiling, editing, or running TURBO PASCAL programs, see the documentation supplied with the program or read one of the many books on this language.

2.5 Running the simulation

If the programs are available on disk, then this simulation can be run by booting the computer in the usual way from either a hard disk or a floppy disk and obtaining the “>” prompt. If the computer was booted from a floppy, then remove this system disk and insert the disk with the Monte Carlo programs on and type *Randomwalk.* at the “>” prompt. Be sure that the disk (if it is a copy) has the .BGI graphics driver files on it or else the program will crash. If the computer was booted from a hard disk, then put the program floppy in the first available drive (usually labeled A:) and type *A:Randomwalk.* If the program has been typed into the computer from the listing, then save it to disk before compiling and running it, using the instruction provided by the relevant manual for the language chosen. If you are planning to make any changes to the program, always work from a copy rather than from an original.

When the welcome screen appears, enter the number of steps required, hit return, and the simulation will run. To generate another run, hit return (to get the “>” prompt once more). If your keyboard has function keys, then hitting F1 will redisplay the original instruction (i.e., “A:Randomwalk”), and hitting Return will cause the program to restart. Otherwise repeat the process as described above. Figure 2.3 shows some typical results for 200-step simulations. Every run will give a different track, and each will give a different distance between the start and finish of the walk. As is evident from Fig. 2.4, which plots distances recorded for several

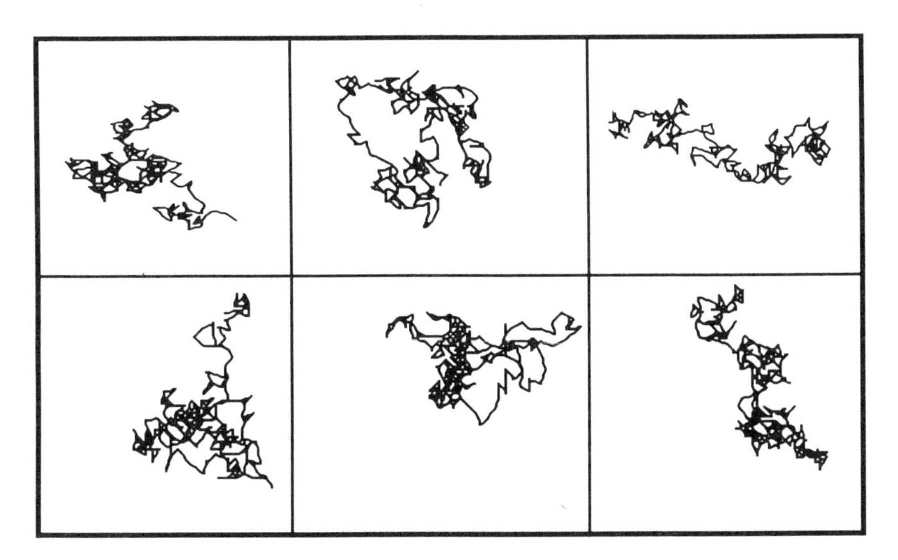

Figure 2.3. Six random walks, each of 200 steps.

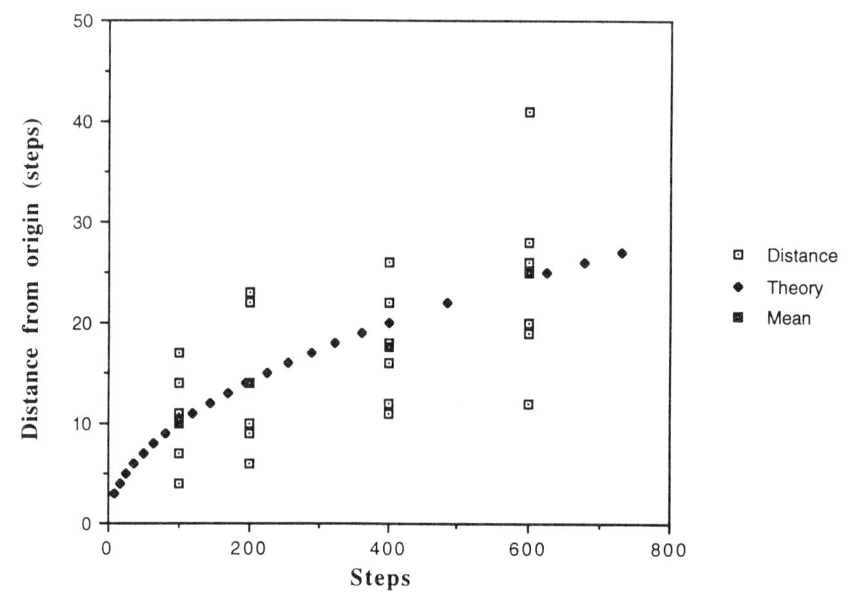

Figure 2.4. Summary of random walk data.

successive runs at a given number of steps, the radial distance traveled from the starting point can vary over a wide range. This is to be expected, because the simulation is, in effect, taking samples of a random walk and there will therefore be a statistical scatter in the data points obtained. If the samples are assumed to be independent, which implies among other things that the random numbers used in the simulation are really random and not in any way related, then the relative precision of the result will be proportional to the inverse of the square root of the number of samples taken. Thus a relative precision of 10% would be achieved on 100 samples, while to reach a 1% precision would require 10,000 samples. So while a single simulation of a random walk of a given number of steps would produce little useful information, a series of such simulations, when averaged, would give us a mean distance traveled and an estimate of the relative precision of this mean value. Thus plotting (see Fig. 2.4) the average value of the distance traveled, determined from all the simulations using the same number of steps, as a function of the number of steps taken gives a series of values that falls close to the predicted theoretical result that the distance traveled is equal to the square root of the number of steps taken. An essential requirement of any Monte Carlo simulation, therefore, is the necessity to repeat the simulation a sufficiently large number of times to obtain a statistically meaningful result.

3

THE SINGLE SCATTERING MODEL

3.1 Introduction

In this chapter we will develop a Monte Carlo model for the interaction of an electron beam with a solid using the "single scattering" approximation. Within the limits discussed below, this model is the most accurate representation of the electron interaction that we can construct, and it is capable of giving excellent results over a wide range of conditions and materials. This program, together with the faster but less detailed "plural scattering" model developed in the following chapter, will form the framework around which specific applications of Monte Carlo modeling to problems in electron microscopy and microanalysis will be generated in subsequent chapters.

3.2 Assumptions of the single scattering model

The single scattering Monte Carlo simulation calculates the passage of the electron through the solid by tracking it from one interaction to the next. In principle, these interactions could be either elastic or inelastic, so the resultant changes in the direction of motion and the energy of the electron would be computed on the basis of the specific sequence of scattering events that each electron suffered. While this is a feasible way of tackling the problem, it has the disadvantage that because not all scattering events are equally important—in terms of their effect on the trajectory or energy—or probable, the amount of computation required becomes excessively large. The procedure described here makes two significant assumptions to provide an accurate simulation while minimizing the amount of calculation required.

1. Only elastic scattering events are considered in determining the path taken by any given electron. Elastic scattering is the net deviation that the incident electron undergoes as a result of the coulombic attraction between the negatively charged electron and the positively charged atomic nucleus and the corresponding repulsion between the orbiting electrons and the incident electron. This process, which is described mathematically by the screened Rutherford cross section, results in angular deflections of from typically 5° to 180°. (For an excellent concise discussion on electron scattering phenomena see Egerton, 1987.) Inelastic scattering, on the other hand, produces deflections of the order $\Delta E/E$ [i.e., the ratio of ΔE, the

energy lost in the event, to E, the energy of the electron (Egerton, 1987)]. Thus a 10-keV electron which deposits 15 eV in an inelastic collision is deflected through an angle of about (15/10,000) radians (i.e., 1.5 milliradians or about 0.1°). An electron that loses more energy in an inelastic event is, of course, deflected through a larger angle, but the probability of these events is proportional to about $1/\Delta E$, so only a small fraction of inelastic events will produce significant deviations of the trajectory. Consequently elastic scattering events are the ones that dominate in determining the spatial distribution of the interaction; ignoring the effects of the inelastic scattering produces little error.

2. The electron is assumed to lose energy continuously along its path at a rate determined by the Bethe relationship rather than as the result of discrete inelastic events. While some energy loss mechanisms are continuous, such as the production of *Bremsstrahlung* ("braking radiation") by the slowing down of the incident electron as a consequence of electron-electron interactions, most energy losses are associated with a specific event such as the production of an inner shell ionization. However, the averaging of all such effects along the path taken by the electrons leads to a convenient simplification in which all possible types of inelastic event are taken account of by one simple expression involving only a single variable. This model of continuous energy loss is therefore in almost universal use—although, as discussed later, significant modifications to its functional form may become necessary at low beam energies.

3.3 The single scattering model

The basic geometry for the simulation (Fig. 3.1) assumes that the electron undergoes an elastic scattering event at some point P_n, having traveled to P_n from a previous scattering event at P_{n-1}. The fundamental task of the simulation is to compute the coordinates of the point P_{n+1} to which the electron travels as a result of the scattering event at P_n. The parameters that describe the instantaneous situation of the electron are its energy E and the direction cosines of the trajectory segment that brought it from P_{n-1} to P_n. The direction cosines cx, cy, cz are defined in the same way as the direction cosines employed in the random walk simulation except that in this three-dimensional situation there are now three rather than two axes to be considered. The X, Y, and Z axes in all the simulations in this book are defined with the convention that the positive Z axis is normal to the specimen surface and directed into the specimen, the X axis is parallel to the tilt axis, and the X-Y plane is the surface plane of the untilted sample. When the sample is tilted, the positive direction of the Y axis is down the surface of the specimen.

To calculate the position of the new scattering point P_{n+1}, we first require to know the distance, or "step," between P_{n+1} and the preceding point P_n. As discussed above, an assumption of this particular model is that only elastic scattering events are explicitly considered. The distance between successive scattering events

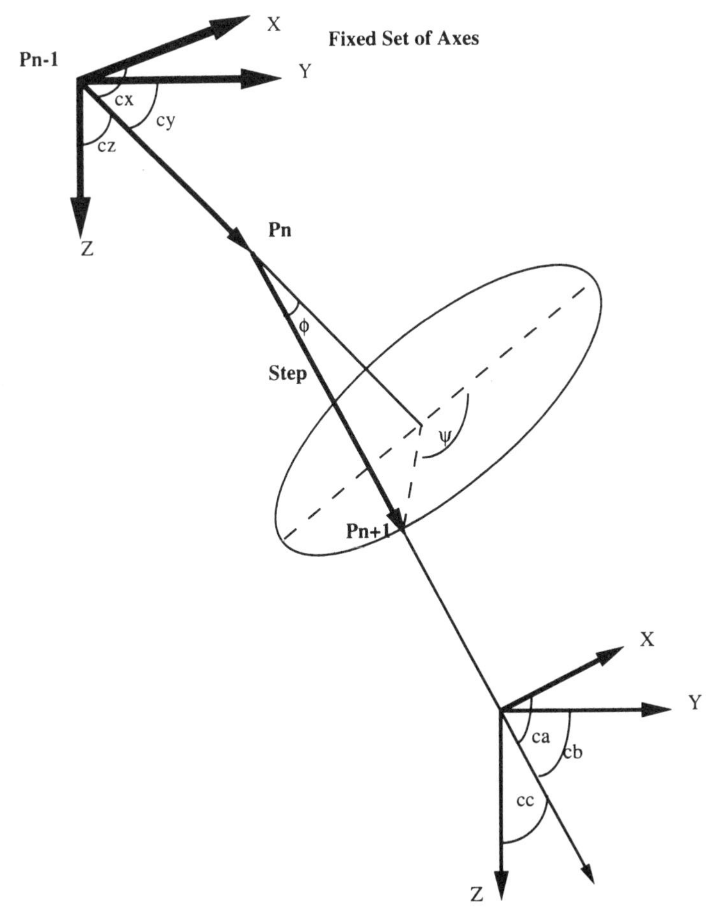

Figure 3.1. Definition of coordinate system used in the Monte Carlo simulation program in this chapter.

is therefore related to the elastic mean free path—which, in turn, is a function of the elastic cross section σ_E. The total screened Rutherford elastic cross section σ_E is given by the relation (Newbury and Myklebust, 1981):

$$\sigma_E = 5.21 \times 10^{-21} \frac{Z^2}{E^2} \frac{4\pi}{\alpha(1 + \alpha)} \left(\frac{E + 511}{E + 1024} \right)^2 \quad \text{cm}^2/\text{atom} \tag{3.1}$$

where E is the electron energy in kilo-electron volts, Z is the atomic number of the target, and α is a "screening factor" that accounts for the fact that the incident electron does not see all of the charge on the nucleus because of the cloud of orbiting electrons. α is usually evaluated using an analytical approximation, such as that given by Bishop (1976):

$$\alpha = 3.4 \times 10^{-3} \frac{Z^{0.67}}{E} \tag{3.2}$$

The cross section σ_E defines a mean free path λ, which is given by the formula:

$$\lambda = \frac{A}{N_a \, \rho \, \sigma_E} \quad \text{cm} \tag{3.3}$$

where N_a is Avogadro's number, ρ is the density of the target in g/cm^3, and A is the atomic weight of the target in g/mole. λ represents the average distance that an electron will travel between successive elastic scattering events. Calculating λ from the formulas above for a selection of elements shows (Table 3.1) that λ is typically of the order of a few hundred angstroms at 100 keV and about one-tenth of that at 10 keV. Materials with higher atomic numbers cause more elastic scattering than elements of lower atomic number, therefore they have a shorter mean free path.

The actual distance that an electron travels between successive elastic scatterings will, of course, vary in a random fashion. The probability $p(s)$ of an electron traveling a distance s when the mean free path is λ is

$$p(s) = \exp(-s/\lambda) \tag{3.4}$$

An estimate for the distance actually traveled can then be found by choosing a random number RND and solving Eq. (1.1) in the form

$$\text{RND} = \frac{\int_0^s \exp(-s/\lambda) \, ds}{\int_0^\infty \exp(-s/\lambda) \, ds} \tag{3.5}$$

which gives

$$\text{RND} = (1 - \exp(-s/\lambda) \tag{3.6}$$

Table 3.1 Elastic mean free paths

Element	Z	λ at 100 keV	λ at 10 keV
Carbon	6	1310 Å	170 Å
Silicon	14	1112 Å	127 Å
Copper	29	297 Å	35 Å
Gold	79	89 Å	10 Å

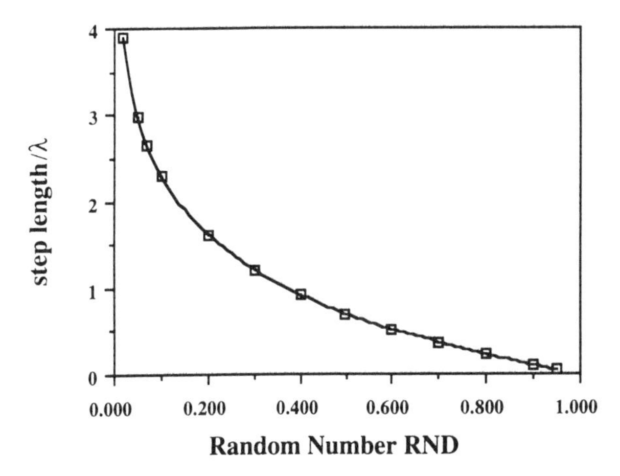

Figure 3.2. Variation of step length, in units of the mean free path, with the random number chosen.

and hence

$$s = -\lambda \log_e(1 - \text{RND}) = -\lambda \log_e(\text{RND}) \tag{3.7}$$

(since RND is a random number between 0 and 1, 1 − RND is also a random number in the same range and so can be replaced by yet another random number RND). Figure 3.2 plots the variation of step length, in units of λ, as a function of the random number chosen, using the result of Eq. (3.7). Since the random numbers are uniformly distributed between 0 and 1, there is, for example, a 10% chance of drawing a number such that RND ≤ 0.1 and so (from Fig. 3.2) of getting a step length such that $s \geq 2.3\ \lambda$. Equally there is a 10% chance of drawing a number such that RND ≥ 0.9, in which case the step length $s \leq 0.1\ \lambda$. The step lengths therefore vary over a wide range of values, depending on the random number picked by the computer. However, as is readily shown from Eq. (3.7), the average value of the step length s resulting from a large number of tries will be λ, as would be expected from the definition of λ as the *average* distance between successive scatterings.

In the scattering event at P_n, which marks the start of the step, the electron is deflected through some angle ϕ relative to its previous direction of travel (see Fig. 3.1). The way in which this scattering occurs is described by σ', the angular differential form of the Rutherford cross section (e.g., Reimer and Krefting, 1976):

$$\sigma' = \frac{d\sigma}{d\Omega} = 5.21 \times 10^{-21} \frac{Z^2}{E^2} \left(\frac{E + 511}{E + 1024}\right)^2 \frac{1}{\left(\sin^2\left(\frac{\phi}{2}\right) + \alpha\right)^2} \tag{3.8}$$

where α is the screening parameter discussed above. An appropriate function to select a scattering angle using a random number can then be obtained from a further application of Eq. (1.1) in the form:

$$\text{RND} = \int_{\Omega} \frac{\sigma'(\phi)}{\sigma_E}\, d\Omega \tag{3.9}$$

This integration, using the result for σ_E given in Eq. (3.1), yields the relation for the scattering angle (Newbury and Myklebust, 1981):

$$\cos\phi = 1 - \frac{2\alpha\ \text{RND}}{(1 + \alpha - \text{RND})} \tag{3.10}$$

This equation generates a unique scattering angle in the range $0 \leq \phi \leq 180°$, producing an angular distribution that matches the one obtained experimentally. Although the full range of angles between 1 and 180° is available, the great majority of scattering events are predicted by Eq. (3.10) to be low-angle—that is, less than about 10°. Figure 3.3 plots the probability of obtaining a scattering angle ϕ greater than some minimum value Φ for the particular case of a silicon target irradiated at 100 keV. Note that while there is only a 1 in 10,000 chance of an electron being scattered by an angle in excess of 110°, more than 50% of all electrons are scattered through at least 1.5°.

The electron can scatter freely to any point on the base of the cone shown in Fig. 3.1. The azimuthal scattering angle ψ is then given as

$$\psi = 2\pi.\ \text{RND} \tag{3.11}$$

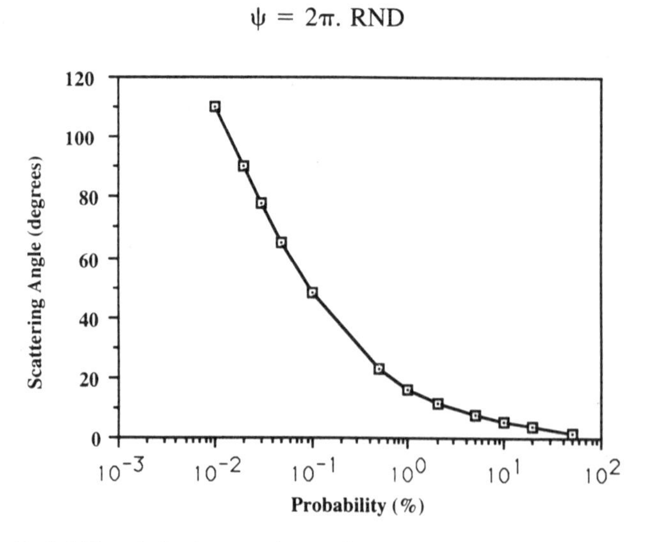

Figure 3.3. Probability of elastic scattering angle exceeding some specified minimum value.

where RND is an independent random number selected by the computer. Since ϕ, ψ, and the step length are now determined, the relationship of P_{n+1} to P_n can now be found. The procedure is similar to that used in Chap. 2, except that the motion is now in three rather than two dimensions. As before, the path is described using direction cosines, *ca*, *cb*, and now *cc*. The coordinates at the end of the step at P_{n+1}, *xn*, *yn*, and *zn*, are then related to the coordinates x, y, z at P_n by the formulas

$$xn = x + \text{step}.ca \quad (3.12a)$$

$$yn = y + \text{step}.cb \quad (3.12b)$$

$$zn = z + \text{step}.cc \quad (3.12c)$$

The direction cosines *ca*, *cb*, *cc* are, as previously, found from the direction cosines *cx*, *cy*, and *cz* with which the electron reached P_n. The derivation of the required transformation from *cx*, *cy*, and *cz* to *ca*, *cb*, and *cc* is similar in outline to that given in Chap. 2 but is more complex because of the extra dimension. The result (modified from the derivation of Myklebust et al., 1976) is

$$ca = (cx.\cos\phi) + (V1.V3) + (cy.V2.V4) \quad (3.13a)$$

$$cb = (cy.\cos\phi) + (V4.(cz.V1 - cx.V2)) \quad (3.13b)$$

$$cc = (cz.\cos\phi) + (V2.V3) - (cy.V1.V4) \quad (3.13c)$$

where

$$V1 = AN.\ \sin\phi \quad (3.14a)$$

$$V2 = AN.AM.\ \sin\phi \quad (3.14b)$$

$$V3 = \cos\psi \quad (3.14c)$$

$$V4 = \sin\psi \quad (3.14d)$$

and

$$AN = -(cx/cz) \quad (3.15a)$$

$$AM = \frac{1}{\sqrt{(1 + AN.AN)}} \quad (3.15b)$$

Note that the sum of the squares of the direction cosines is always equal to unity,

$$cx^2 + cy^2 + cz^2 = ca^2 + cb^2 + cc^2 = 1 \quad (3.16)$$

As the electron travels through the specimen, it loses energy continuously because of the drag exerted on the negatively charged electron by the positively charged nuclei surrounding it; it also loses energy during discrete inelastic scattering events such as the horizon or production of a plasmon. The most complete approach to accounting for these energy losses would be to incorporate them individually into the simulation, but this is a relatively lengthy procedure because of the number of different possibilities available. Instead, we make the assumption that the effect of both the electrostatic drag on the electron and the discrete energy losses occurring during inelastic scattering events can be combined and approximated by a model in which the incident electron is slowing down continuously as it travels. The rate at which energy is lost by the incident electron was shown by Bethe in a classic paper (1930) to be expressible in the form:

$$\frac{dE}{dS} = -78{,}500 * \frac{Z}{AE} * \log_e\left(\frac{1.166E}{J}\right) \quad (3.17)$$

where (dE/dS) is sometimes referred to as the "stopping power" of the target, E is the energy of electron (in kilo-electron volts) Z and A are respectively the atomic number and atomic weight of the target, and S is the product of ρ the density of the target (in g/cm^3) and the distance traveled along the trajectory s. J, which has units of kilo-electron volts, is called *the mean ionization potential* and represents the effective average energy loss per interaction between the incident electron and the solid. This single parameter incorporates into its value all possible mechanisms for energy loss that the electron can encounter, thus allowing the Bethe equation to provide a convenient and compact way of accounting for the variety of energy losses experienced.

J has been measured experimentally, using nuclear physics techniques, for a wide range of materials and compounds (see for example ICRU, 1983). Table 3.2 gives values for some common elements (in solid form), and it can be seen that, in general, there is a monotonic and almost linear increase of J with the atomic number of the element. Berger and Seltzer (1964) showed that this variation could be fitted with good accuracy by the relation

$$J = \left[9.76\,Z + \frac{58.5}{Z^{0.19}}\right] . \ 10^{-3} \quad (\text{keV}) \quad (3.18)$$

and a comparison of this expression with the data in Table 3.2, for example, for silicon ($Z = 14$) and gold ($Z = 79$), shows that the fitted value and the experimental value are

Table 3.2 Measured values of mean ionization potential J

Element	J (eV)	Element	J (eV)	Element	J (eV)
H	21.8	Cr	257	Pd	470
Li	40.0	Mn	272	Ag	470
Be	63.7	Fe	286	Cd	469
B	76.0	Co	297	Sn	488
C	78.0	Ni	311	Gd	591
Na	149	Cu	322	Ta	718
Al	166	Zn	330	W	724
Si	173	Ga	334	Pt	790
Ca	191	Ge	350	Au	790
Sc	216	Zr	393	Pb	823
Ti	233	Nb	417	U	890
V	245	Mo	424		

within a few electron volts of each other. For compounds, the appropriate value of J can again be found using Eq. (3.18), but replacing Z by Z_{av}, the mean value of Z for the compound. So, for example, if a material has the composition AB_2 (i.e., 33 atomic % of A and 66% of B), then

$$Z_{av} = (1^*Z_A + 2^*Z_B)/3 \tag{3.19}$$

where Z_A and Z_B are the atomic weights of A and B respectively. This simple average in most cases produces a value for Z_{av}, and thus for J, which is of acceptable accuracy. In the case of complex materials, however, the composition may not be known. Table 3.3 therefore gives J values, again derived from the ICRU (1983) report, for some compounds likely to be encountered in electron microscopy. Although this list is far from exhaustive, it does give a useful guide as to probable values for generic types of materials.

Table 3.3 Measured values of mean ionization potential J

Material	J (eV)	Material	J (eV)	Material	J (eV)
Nylon	63.9	Aluminum oxide	145	Adipose tissue	63
Teflon	99.1	Calcium fluoride	166	Bone	107
Paraffin wax	48.3	Lithium fluoride	94	Muscle	75
PMMA	74.0	Silicon dioxide	139	Skin	74
Polyethylene	57.4	Sodium iodide	452	Air (1 atm.)	86
Polystyrene	68.7	"Pyrex" glass	134	Blood	75
Plastic scintillator	64.7	Photographic emulsion	64.7	Liquid water	75

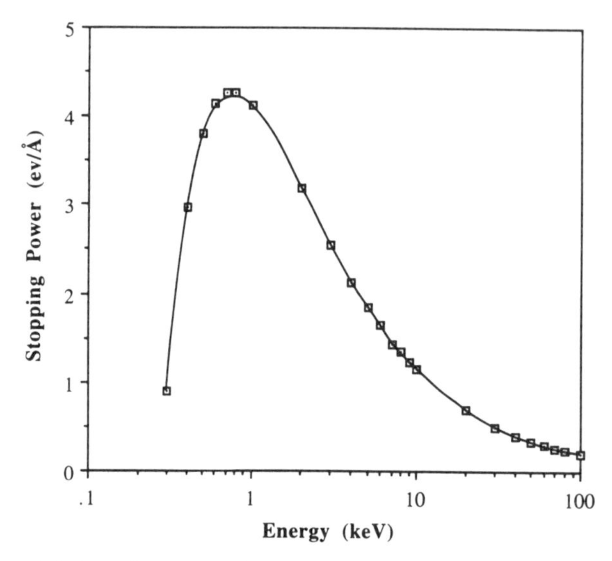

Figure 3.4. Stopping power for copper computed from Bethe expression.

Using either the measured or computed value of J, Eq. (3.17) can now be evaluated for a given electron energy E. Figure 3.4 attempts to plot the rate of energy loss, $(-dE/ds)$, or $\rho(dE/dS)$, for copper ($\rho = 7.6$ g/cm^3) over the energy range 100 eV to 100 keV. The predicted energy-loss values show a steady increase from about 0.1 eV per angstrom at 100 keV up to almost 10 eV per angstrom at beam energies of a few kilo-electron volts. These values are typical of those found in all materials. However at an energy of about 2 keV, the curve reaches a maximum value before starting to fall rapidly as the energy is further reduced. This behavior is obviously not physically realistic but comes from the change in sign of the logarithmic term in Eq. (3.17) when $E \leq J$. Therefore, while the Bethe stopping-power formula is an excellent approximation for energies such that $E \gg J$, it cannot be used at lower energies without encountering problems. In particular, since J is of the order of 0.3 to 0.6 keV for most materials, the simulation of the trajectory of electrons traveling with initial energies of 1 to 10 keV (i.e., typical energies encountered in a scanning electron microscope) will be difficult, since a significant part of the trajectory will occur in the energy range where the Bethe relation cannot be applied. This problem can be overcome with a method due to Rao-Sahib and Wittry (1974). This is a parabolic extrapolation from the tangent to the Bethe curve at the energy $E = 6.4\,J$, where the curve has an inflection, down to $E = 0$. Thus for $E < 6.4\,J$, Eq. (3.17) is replaced by the expression:

$$\frac{dE}{dS} = -\frac{62400\,Z}{\sqrt{E\,J}\,A} \qquad (3.20)$$

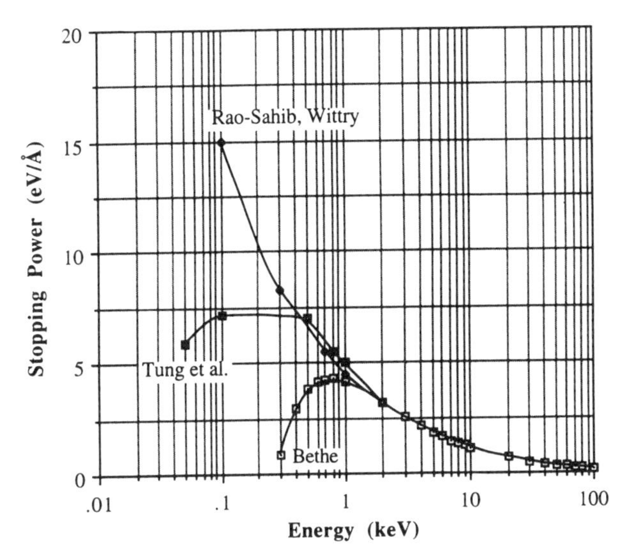

Figure 3.5. Comparison of stopping power for copper from Bethe expression, Rao-Sahib and Wittry extrapolation, and Tung et al. calculation.

This expression provides a convenient extrapolation that is well behaved over the low energy range, and so this approximation has been widely used in Monte Carlo simulations. It is not, however, either physically realistic or accurate. Figure 3.5 compares the rate of energy loss, $(-dE/ds)$, or $\rho(dE/dS)$, for copper ($\rho = 7.6$ g/cm^3) over the energy range 100 eV to 100 keV, obtained using the Bethe model and the Rao-Sahib–Wittry extrapolation, with the results of detailed computations by Tung et al. (1979). The agreement between the composite Bethe-Rao-Sahib-Wittry profile and the Tung et al. data is satisfactory at energies down to a few kilo-electron volts. However, once the energy falls below 1 keV, neither the original Bethe curve nor the Rao-Sahib–Wittry extrapolation is close to the Tung et al. value, the error being a factor of two to three times at an energy of a few hundred electron volts. The same sort of result is found for all other elements, with the errors becoming worse for high atomic numbers, since these have a larger J value [Eq. (3.18)]; consequently the Bethe expression becomes unusable at energies as high as a 4 or 5 keV.

We can simultaneously escape the mathematical singularity of the original Bethe expression, eliminate the need to use the Rao-Sahib–Wittry extrapolation, and improve the accuracy of the stopping power as compared to the Tung et al. data by using a modified version of the Bethe equation suggested by Joy and Luo (1989). We write the stopping power as:

$$\frac{dE}{dS} = -78{,}500 * \frac{Z}{AE} * \log_e\left(\frac{1.166\,(E + 0.85\,J)}{J}\right) \tag{3.21}$$

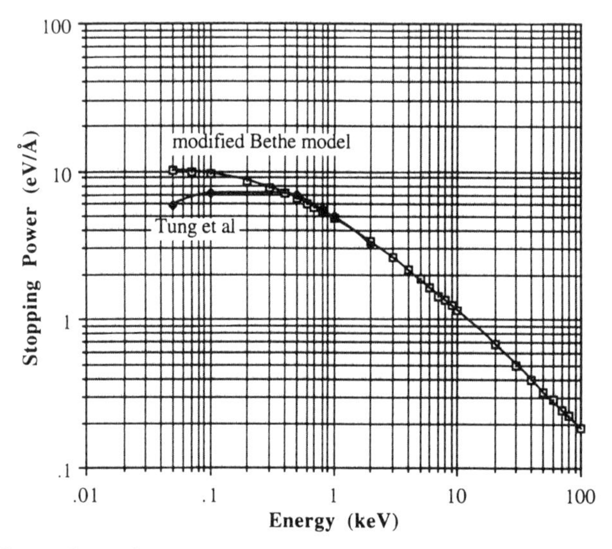

Figure 3.6. Comparison of stopping power for copper from modified Bethe law, with data from Tung et al.

At high energies $E \gg J$, this expression converges to the original Bethe expression of Eq. (3.17) but the addition of the extra term 0.85 J in the numerator of the logarithmic term ensures that for all positive values of E the log term evaluates to a positive quantity. As shown in Fig. 3.6, this modified expression is now a very good fit to the Tung et al. data for all energies above 50 eV. It can be shown that this result is equally applicable to all other elements and compounds of interest, producing a stopping power value which agrees well with more detailed calculations for all energies above a few tens of electron volts. This expression [Eq. (3.21)] is therefore used in this book.

The sequence of operations needed to simulate the electron trajectory through the specimen can now be written schematically in an algorithmic form. Conventionally the electron is allowed to penetrate the first step length into the sample before being scattered. Thus the procedure is:

Calculate initial entry depth into sample [Eq. (3.7)]

then repeat

Get starting energy E of the electron

Determine initial coordinates x, y, z for this step

Find the direction cosines cx, cy, cz of its motion relative to the axes

Compute the mean free path λ for this material and energy E [Eq. (3.3)]

Calculate the step length [Eq. (3.7)]

Find the scattering angles ϕ, ψ from Eqs. (3.10) and (3.11)

Find finish coordinates *xn, yn, zn* for this step [Eqs. (3.12) to (3.15)]
Compute final energy for this step $E' = E -$ step*ρ*(dE/ds)
Reset coordinates $x = xn$, $y = yn$, $z = zn$
Reset direction cosines $cx = ca$, $cy = cb$, $cz = cc$
Reset energy $E = E'$
until the electron leaves the sample or falls below some minimum energy

The PASCAL code that implements this outline is set out below, using the conventions introduced in the previous chapter. Note that, by design, the program eliminates pleasing but unessential features like the use of color, or windows, so as to keep the code as clear and easy to follow as possible.

3.4 The single scattering Monte Carlo code

```
Program SingleScatter;

   {a single scattering Monte Carlo simulation which uses the
    screened Rutherford cross section}

   {$N+}    {turn on numeric coprocessor}
   {$E+}    {turn on emulator package}

uses CRT,DOS,GRAPH;    {resources required}

label    exit;

const    two_pi=6.28318;    {2π}
         e_min=0.5;         {cutoff energy in keV in bulk case}
var
   at_num,at_wht,density,inc_energy,mn_ion_pot:extended;
   al_a,bk_sct,cp,er,ga,lam_a,sp,sg_a,:extended;
   ca,cb,cc,cx,cy,cz,x,y,z,xn,yn,zn:extended;
   del_E,m_step,m_t_step,s_en,step:extended;
   hplot_scale,m_f_p,plot_scale,thick:extended;
   count,k,num,traj_num:integer;
   bottom,center,top:integer;
   thin:Boolean;
   GraphDriver:Integer;
   GRAPHMODE:Integer;
   ErrorCode:Integer;
   Xasp,Yasp:word;

Function power(mantissa,exponent:real):real;
   {because PASCAL does not have an exponentiation function}
```

```
  begin
    if mantissa<=0 then power:=0
  else
    power:=exp(ln(mantissa)*exponent);
  end;

Function stop_pwr(energy:real):real;
    {this computes the stopping power in keV/g/cm2 using the
     modified Bethe expression of Eq. (3.21)}

var temp:extended;
  begin
    if energy<0.05 then energy:=0.05; {just in case}
     temp:=ln(1.166*(energy+0.85*mn_ion_pot)/mn_ion_pot);
       stop_pwr:=temp*78500*at_num/(at_wht*energy);
  end;

Function lambda(energy:real):real;
    {computes elastic MFP for single scattering model}
var al,ak,sg:real;
  begin
    al:=al_a/energy;
    ak:=al*(1.+al);
    {giving sg cross-section in cm2 as}
    sg:=sg_a/(energy*energy*ak);
    {and lambda in angstroms is}
    lambda:=lam_a/sg;
  end;

Function yes:Boolean;
    {reads the keyboard for 'y' or 'Y'}
var ch:char;
  begin
    read(ch);
     if ch in ['Y', 'y'] then yes:=true
      else
     yes:=false;
  end;

Procedure get_constants;
    {computes some constants needed by the program}
  begin
              al_a:=power(at_num,0.67)*3.43E-3;
      {relativistically correct the beam energy for use up to 500 keV}
               er:=(inc_energy+511.0)/(inc_energy+1022.0);
                 er:=er*er;
               lam_a:=at_wht/(density*6.0E23);    {lambda in cm}
```

```
            lam_a:=lam_a*1.0E8;                {put into angstroms}
            sg_a:=at_num*at_num*12.56*5.21E-21*er;
 end;

Procedure set_up_screen;
   {gets the input data to run this program}

 begin
     ClrScr; {erases all previous data from screen}
     GoToXY(22,1);
       Writeln ('Single Scattering Monte Carlo Simulation');

  {Having set up the screen, now get input data}

    GoToXY(1,5);
      Write ('Input beam energy in keV');
        Readln(Inc_Energy);

    GoToXY(1,7);
      Write('Target Atomic Number is');
        Readln(at_num);

   {Calculate J the mean ionization potential mn_ion_pot using the
    Berger-Selzer analytical fit}

   mn_ion_pot:=(9.76*at_num + (58.5/power(at_num,0.19)))*0.001;
 GoToXY(1,9);
      Write('Target Atomic Weight is');
         Readln(At_wht);

    GoToXY(1,11);
      Write('Target density in g/cc is');
         Readln(Density);

    GoToXY(40,5);
        write('Is this a bulk specimen (Y/N)?');

        if yes then     {it's thick}
        begin           {so estimate the beam range for graphics scale}
          thick:=700.0*power(inc_energy,1.66)/density;
             if thick<1000.0 then thick:=1000.0;
          thin:=false;
         end

        else            {it's thin}
         begin
             GoToXY(40,7);
```

```
              write('Foil thickness (A)');
               readln(thick);
                 thin:=true;
        end;

   {get the number of trajectories to be run in this simulation}
         GoToXY(40,9);
          write('Number of trajectories required');
            readln(traj_num);

  end;

Procedure initialize;
   {sets up the graphics drivers for version 5.0 TURBO PASCAL}

var
   InGraphicsMode:Boolean;
   PathToDriver:String;

  begin

   DirectVideo:=False;
    PathToDriver:='';
     GraphDriver:=detect;
    InitGraph(GraphDriver,GRAPHMODE,PathToDriver);
   SetViewPort(0,0,GetMaxX,GetMaxY,True); {clips display area}
             center:=trunc(GetMaxX/2);    {beam entry coordinate}
             top:=trunc(GetMaxY*0.1);     {beam entry coordinate}
  end;

Procedure xyplot(a,b,c,d:real);
   {this displays the trajectories on the pixel screen}

var iy,iz,iyn,izn:integer;

  begin
    iy:=center + trunc(a*hplot_scale);     {plotting coordinate #1}
    iz:=top + trunc(b*plot_scale);         {plotting coordinate #2}
    iyn:=center + trunc(c*hplot_scale);    {plotting coordinate #3}
    izn:=top + trunc(d*plot_scale);        {plotting coordinate #4}
          if d=99 then izn:=top-2;         {BS plotting limit}
          if d=999 then izn:=bottom+2;     {transmitted plotting
                                           limit}
    {and now plot this vector on the screen}
           line(iy,iz,iyn,izn);
  end;
```

```
Procedure set_up_graphics;
   {draws in the surface(s), beam location, and action thermometers}

var
   a,b:integer;

  begin
                  a:=GetMaxX-20;      {adjust to suit your screen}
                  b:=20;              {ditto}

            Line(b,top,a,top);        {plot in top surface}
           {find position of bottom surface}
             bottom:=top + trunc(thick*GetMaxY/1000);
                 if bottom>GetMaxY-40 then bottom:=GetmaxY-40;

         plot_scale:=(bottom-top)/thick; {hence get pixels/angstrom}
           GetAspectRatio(Xasp,Yasp);
         hplot_scale:=(Yasp/Xasp)*plot_scale; {for aspect ratio}

         if thin then {also plot in exit surface}
             Line(b,bottom,a,bottom);      {plot bottom surface}
             Line(center,1,center,top);    {plot incident beam}
         if not thin then {put up micron markers}
          if thick>5000 then
           begin
            Line(b,bottom+10,trunc(b+5000*plot_scale),bottom+10);
            OutTextXY(b,bottom+20,'0.5 microns');
           end
         else
           begin
            Line(b,bottom+10,trunc(b+500*plot_scale),bottom+10);
            OutTextXY(b,bottom+20 '500A');
           end;

   {put up a thermometer for the trajectories completed}

OutTextXY(trunc(center-80),bottom+15'0% . . . . . . . 50% . . . . . .
            100%');
            OutTextXY(trunc(center-78),bottom+28'Trajectories
            completed');
  end;

Procedure reset_coordinates;
   {resets coordinates at start of each trajectory}

  begin
```

```
      s_en:=inc_energy;
       x:=0;
        y:=0;
         z:=0;
        cx:=0;
       cy:=0;
      cz:=1;
  end;

Procedure zero_counters;
   {since PASCAL does not zero variables on start-up we must do this}
  begin
        bk_sct:=0;
          num:=0;
  end;

Procedure s_scatter(energy:real);
   {calculates scattering angle using screened Rutherford cross
section}

var R1,al:real;

  begin
       al:=al_a/energy;
        R1:=random;
        cp:=1.0-((2.0*al*Rq)/(1.0+al-R1));
        sp:=sqrt(1.0-cp*cp);
  {and get the azimuthal scattering angle}
        ga:=two_pi*random;
  end;

Procedure new_coord(step:real);
   {gets xn,yn,zn from x,y,z and scattering angles}

var an_n,an_m,v1,v2,v3,v4:real;
  begin
   {find the transformation angles}
                   if cz=0 then cz:=0.000001;
                      an_m:=(-cx/cz);
                      an_n:=1.0/sqrt(1+(an_m*an_m));
  {save computation time by getting all the transcendentals first}
                      v1:=an_n*sp;
                       v2:=an_m*an_n*sp;
                       v3:=cos(ga);
                      v4:=sin(ga);
  {find the new direction cosines}
                      ca:=(cx*cp) + (v1*v3) + (cy*v2*v4);
```

```
                    cb:=(cy*cp) + (v4*(cz*v1 - cx*v2));
                    cc:=(cz*cp) + (v2*v3) - (cy*v1*v4);
 {and get the new coordinates}
                    xn:=x + step*ca;
                    yn:=y + step*cb;
                    zn:=z + step*cc;
 end;

Procedure straight_through;
   {handles case where initial entry exceeds thickness}

 begin
     num:=num+1;
     xyplot (0,0,0,999);
 end;

Procedure back_scatter;
   {handles case of backscattered electrons}
 begin
    num:=num+1;
    bk_sct:=bk_sct+1;
    xyplot (y,z,yn,99);
 end;

Procedure transmit_electron;
   {handles case of transmitted electron}

var l1:extended;
 begin
        num:=num+1;
        l1:=(thick-z)/cc;   {length of path from z to bottom face}
        yn:=y+l1*cb;   {hence the exit y-coordinate}
        xyplot(y,z,yn,999);
 end;

Procedure reset_next_step;
   {resets variables for next trajectory step}
 begin
      xyplot(y,z,yn,zn);
       cx:=ca;
        cy:=cb;
         cz:=cc;
        x:=xn;
       y:=yn;
      z:=zn;
     {find the energy lost on this step}
       del_E:=step*stop_pwr(s_en)*density*1E-8;
```

```
     {so the current energy is}
        s_en:=s_en - del_E;
  end;

Procedure show_traj_num;
   {updates thermometer display for % of trajectories done}

var   a,b:integer;
  begin
                a:=trunc(center-80);    {adjust to suit your screen}
                b:=bottom+23;           {ditto}

        line(a,b,a+trunc(165*(num/traj_num)),b);

  end;

Procedure show_BS_coeff;
   {displays BS coefficient on thermometer scale}

label hang;

var   a,b:integer;

  begin
                     a:=GetMaxX-180;    {adjust to suit your
                     screen}
                     b:=bottom+23;    {ditto}

              OutTextXY(a,b-8,'0 . . . . 0.25 . . . 0.5 . . . 0.75');
              OutTextXY(a+5,b+5,'BS coefficient');
          Line(a,b,trunc(a+(bk_sct/traj_num)*220),b);

hang:   {this loop freeze the display on the screen }
    If (not keypressed) then goto hang;

CloseGraph; {shut down graphics driver}

  end;

{***************************************************
 *       this is the start of the main program       *
 ***************************************************}

begin
                set_up_screen; {get input data and find J value}
```

```
                get_constants; {and parameters needed later}

          { reset the random number generator}
                 randomize;

     {set up the graphics display for plotting}
                   Initialize;
                 set_up_graphics;


{**********************************************
 *                the Monte Carlo Loop                *
 **********************************************}


                 zero_counters;

              while num < traj_num do
 begin
                  reset_coordinates;

              {allow initial entrance of electron}
                 step:=-lambda(s_en)*ln(random);
                   zn:=step;

            if zn>thick then {this one is transmitted}
              begin
                   straight_through;
                   goto exit;
              end
            else   {plot this position and reset coordinates}
              begin
                   xyplot(0,0,0,zn);
                    y:=0;
                    z:=zn;
              end;

       {now start the single scattering loop}
 repeat
              step:=-lambda(s_en)*ln(random);
                s_scatter(s_en);
                new_coord(step);

            {problem-specific code will go here}

                {decide what happens next}
```

```
                if zn<=0 then {this one is backscattered}
                 begin
                  back_scatter;
                  goto exit;
                 end;

                if zn>=thick then {this one is transmitted}
                 begin
                  transmit_electron;
                  goto exit;
                 end;

               {otherwise we go round again}
                  reset_next_step;

until s_en<=e_min; {end of repeat loop - the energy drops below e
                    _min}

                    num:=num+1;     {increment counter}
          exit:                     {end of goto jumps}

{*********************************************
 *          end of the Monte Carlo Loop       *
 *********************************************}

               show_traj_num;

          if num mod 100=0 then randomize; {reset generator}
   end;

                 show_BS_coeff;

end.
```

3.5 Notes on the procedures and functions used in the program

The program starts with two special statements, {$N+} and {$E+}, which are known as *pragmas* or *compiler directives*. These tell the TURBO PASCAL compiler to choose certain options when it runs the program. The first pragma instructs the compiler to generate the code necessary to make use of a numerical coprocessor chip (for example, an 80387 device in a personal computer running MS-DOS). While computers can rapidly perform arithmetic operations on integers, they are typically slower by a factor of fifty or more times when manipulating real or floating-point numbers (i.e., numbers such as 10.23 or $1.5*10^{-3}$). Floatingpoint coprocessor chips, such as the 80387, are optimized to perform arithmetical operations on real numbers and greatly speed up the computation. However, in order to use the chip, the compiler must be instructed to generate the necessary special code, hence the pragma {$N+}. Unfortunately, code produced in this way would not then run on another machine without the math coprocessor. The second pragma {$E+} tells to compiler to test the computer at the start of the program to determine if a math coprocessor is present. If it is, then the chip is used; if no coprocessor is detected, then the compiler "emulates" the chip in software, using the same code instructions and performing the same operations as the chip, although much more slowly. Together, these two instructions allow the program to configure itself so as to run without modification on any suitable machine. Since Monte Carlo programs depend exclusively on floating-point calculations, a math coprocessor chip is the cheapest and most efficient way of upgrading the performance of the computer.

As in the simpler example given in Chap. 2, the program then continues with a declaration of the resources required. Then `LABEL` collects the names of all of the labels used in the program. A label is the target of a "GoTo" instruction, which causes the execution of the program to jump to the specified location. Finally, we give a definition of any constants used. Here the constants are two_pi (2π), and e_min the minimum energy below which the electron will no longer be tracked. Defining as constants quantities, which are often used, saves time when the program is running because the computer does not have to initialize each quantity many times. Next comes the list of variables used. Those declared in this list are available to all parts of the program and so are called *global* variables, while those declared inside a procedure or function are *private* and only available to that procedure.

The first function called, `POWER`, illustrates how functions can be used to extend the PASCAL language itself. Most versions of the language do not include an operator to find the value of the quantity x^y; the function of POWER therefore provides this ability. Note that the simple code supplied here does not contain any tests to avoid possible problems, such as attempting to take the logarithm of a zero or negative number. A version of this function with proper error trapping would therefore be necessary if the function were to be used in a more general context. The functions `mn_ion_pot`, `stop_pwr`, and `lambda` calculate the quantities suggested by their names, using the equations discussed earlier in this chapter. In each case the function is supplied with a parameter, such as the atomic number of the sample or the electron energy, in order to compute the appropriate value. The other data needed by the functions is taken from the global variables declared at the start of the program. The final function YES waits for input from the keyboard. If this input is a "y," or a "Y," then the answer to the question is assumed to be yes and a Boolean variable is set to true. If the input is any other letter or number, the variable is set to false. In PASCAL, the code statement:

```
If (function) then (operation)
```

will lead to the operation being performed if the function evaluates to a positive number or produces a Boolean variable with the value "true" (i.e., +1). If, on the other hand, the function evaluates to zero or to a negative number or produces a Boolean variable with the value "false" (i.e., 0), then the operation is not performed and the program moves to the next statement. Thus when the program at some later point encounters code such as:

```
If yes then . . . . . . . .
        else . . . . . . . . .
```

the function yes is called and the program waits until input from the keyboard is received, If this input is a "y" or "Y" then yes is set to true and the "if" statement will be performed. Otherwise the program will move on to the next line.

The first procedure `get_constants` computes a variety of quantities needed for other parts of the program, including the relativistic correction to the incident beam energy, and preconstants in the elastic cross section. Next `set_up_screen` allows for the entry of the parameters describing the specimen and experimental conditions. The operator is asked if the sample is bulk (that is, one not sufficiently thin to be electron-transparent). If the answer, as determined by the YES function discussed above, is yes then thick, the range of the electrons in the specimen is estimated from a simple analytical approximation:

$$\text{Range (Å)} = \frac{700 * \text{incident energy (keV)}^{1.66}}{\text{density (g/cm}^3\text{)}} \tag{3.22}$$

and this is later used to set up the scale for plotting the trajectories. A Boolean variable `thin` is also set to false. If, on the other hand, the operator says that the specimen is thin, then the Boolean variable is set to true, and thick is set equal to the actual thickness and used to set up the plotting scale. This is also the thickness against which the position of the incident electron will be tested to determine whether or not it has been transmitted through the foil. Finally, the number of trajectories to be run in the simulation is obtained.

The next procedure, `initialize,` is identical to that used in the previous chapter. As before, it identifies the type of graphics display card fitted to the computer, initializes the card ready for use, determines the plotting size of the screen `(GetMaxX, GetMaxY),` and sets itself up to clip any image features lying off the screen. At the same time two variables—center the horizontal midpoint of the screen and top the position at which to draw the entrance surface of the specimen—are defined. The procedure `xyplot` is similar to that used in the previous chapter. Here we use separate plotting scales `plot_scale` and `hplot_scale` to set up the aspect ratio. We also check for two special values of the last plotting parameter passed to the procedure. If $d = 99$, then the electron has been backscattered and the plotting line is terminated 2 pixels above the entrance surface; if $d = 999$ then the electron has been transmitted through the specimen and the plotting line is finished 2 pixels below the exit surface. These steps help keep the display looking tidy.

`set_up_graphics` is the procedure which draws the entrance surface of the specimen and marks the entrance position of the electron beam. If the Boolean variable `thin` is true, then an exit surface is also drawn at a spacing from the top surface which is proportional to the foil thickness. If this exceeds 1000 Å, then the bottom surface is drawn at 40 pixels from the bottom edge of the screen. The plotting scale `plot_scale` in pixels/angstroms is then determined and the aspect ratio of the screen is found to give a horizontal plotting scale `hplot_scale` such that equal distances in either direction will plot as lines of equal length. No scale marker is needed for a thin foil, since the specified thickness of the specimen provides a built-in scale, but for bulk samples a micron marker must be provided. The program tests the value of the estimated electron range `thick` to decide what length the marker should be. Finally we draw in, using the `OutTextXY` procedure, a "thermometer" scale labeled from 0% to 100%, which will be used to monitor the progress of the simulation. While this is in no way essential to the program, it does add a nice touch to the display. Note that in order to make the same code run equally well on all types of graphics systems, we define all positions on the screen in terms of the size `(GetMaxX, GetMaxY)` of the screen, using two local variables `a` and `b`. As before, however, `a` and `b` may be varied, if desired, to customize the display to individual taste.

Since PASCAL does not initialize variables when it starts running a program, the procedure `zero_counters` does this for the quantities `bk_sct` and `num`, which count the number of backscattered electrons and the total number of electrons respectively. Similarly `reset_coordinates` ensures that, at the start of each new trajectory, all of the conditions are properly initialized, and `zero_counters` resets each of the counters used to check the number of electrons run and the number backscattered. The procedure `s_scatter` calculates the scattering angles, using the relationships given in Eqs. (3.10) and (3.11). These values, the starting electron coordinates `x,y,z`, and the step length are then used by `new_coord` to determine the final coordinates `xn,yn,zn` as well as the direction cosines `ca,cb,cc` relative to the fixed axis system. It is at this point in the program that code specific to the task of interest will usually be inserted, since all of the information about the previous and present position and energy of the electron are available. Once this has been handled, then the program calls `reset_next_step`, which plots the trajectory step, determines the amount of energy lost by the electron as it traveled this step, and equates the now current coordinates, direction cosines, and energy with the starting values for the next step. The final procedures `straight_through`, `transmit_electron`, `back_scatter`, and `show_BS_coeff` handle the various ways in which the electron can exit from the specimen, call the plot routine with the appropriate parameters, and count the events. `show_traj_num` draws a line in the thermometer representing the percentage of trajectories completed, and the procedure `show_BS_coeff` uses very similar code to draw a thermometer representing the percentage of electrons backscattered. In each case local variables `a` and `b` can be altered to adjust the display to personal preference.

Because of all the work handled by the procedures and functions, the actual program section, starting at **begin** and finishing at **end,** is short and simple. After the initial setup, the Monte Carlo loop is entered and the electron is allowed to penetrate into the sample by some distance equal to the step length [generated by applying Eq. (3.7)]. If this distance is greater than the thickness of the sample, then the event is plotted and counted and the loop started again with a new electron. Otherwise the position of the electron is plotted and then, at the label `repeat,` a second loop is started in which the electron is scattered, its new position is calculated, and its energy loss computed to give its new energy. The position of the electron is then tested to see whether it has been either transmitted through the foil or backscattered from it, in which case the program leaves the loop and goes to the appropriate procedure to handle this event before commencing a new trajectory. Finally, the instantaneous energy s_en of the electron is checked by the `repeat . . . until s_en<=e_min` statement to see whether or not the energy is above the cutoff value `e_min.` If it is not, then the loop is exited, the electron is counted, and a new trajectory is started. Otherwise the loop is repeated. This sequence continues until a total number of trajectories equal to `traj_num` have been calculated. When the full number of trajectories has been completed, the program displays the fractional yield of backscattered electrons and then terminates.

3.6 Running the program

The program can be run directly from the disk, if this is available, or typed in and compiled. In addition the .BGI files provided by Borland must be present for the graphics routines to operate. If a printed copy of the screen display is required when running MS-DOS 5.0 or higher, then at the ">" prompt type "graphics" and hit the return key. Anything appearing on the screen can now be transferred to the printer by using the key combination "Control + PrintScr." When the ">" prompt returns, type "SS_MC" and hit return. The computer then requests the parameters required to run the simulation. Values of atomic numbers, atomic weights, and densities for the elements can be found, for example, in Goldstein et al. (1992). For compounds, the usual procedure is to compute an effective atomic number by using Eq. (3.19). The atomic weight is then approximated as being twice the effective atomic number, and the appropriate density value is entered when requested. (Ideally, the procedure would be to use a Monte Carlo method to decide which of the atoms in the compound was being scattered from each time. This method has been used (Murata et al., 1971; Kyser, 1979), but it is not clear that the benefits are worth the added complexity). For some very complex materials and compounds, it may be impossible to derive an effective value for Z, since the chemistry is not known. However, if the substance of interest appears in Table 3.3 [or is listed in the full ICRU (1983) tables], then an inverse procedure can be used. For example, for photographic emulsion, the mean ionization potential J is given in Table 3.3 as 64.7 eV. J is approximately $10.Z_{eff}$ [Eq. (3.18)], where Z_{eff} is the effective atomic number of the material. Hence Z_{eff} is taken as 6.5 and the effective atomic weight as 13. While this is clearly only an approximation, in practice it produces sensible and useful results.

To illustrate the application of this program, let us first consider the case of a high-energy electron beam and a specimen that is thin enough to be electron-transparent (i.e. the situation found in the transmission or scanning transmission electron microscope). Figure 3.7 shows the trajectories calculated for a 100-keV beam and 1000-Å thick foils of carbon, copper, silver, and gold. In each case, 250 trajectories were run, and in response to the question `"Is the sample bulk (y/n)?"` the answer was "n." As the atomic number and the density of the foil

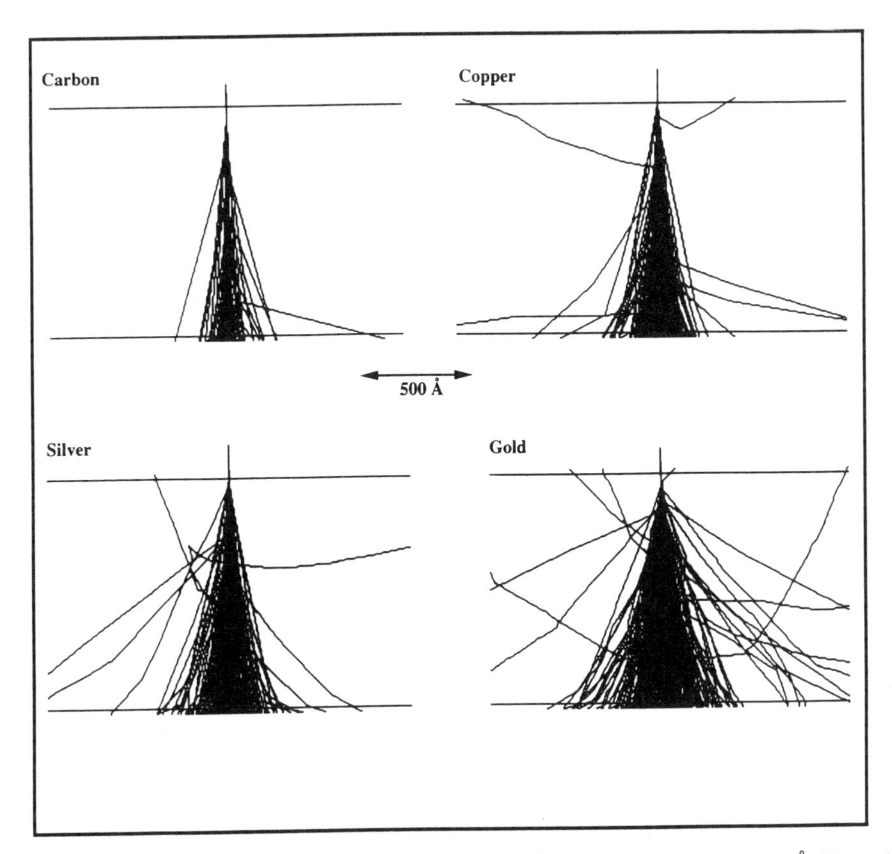

Figure 3.7. Monte Carlo trajectory plots for 100-keV electrons incident on 1000-Å films of carbon, copper, silver, and gold. Number of trajectories per plot: 250.

increase, the amount of lateral scattering in the beam can also be seen to increase. In the case of the carbon sample, the trajectories form a smooth cone that is closely confined around the incident beam axis, and it can be seen from inspection while the program is running that the average electron is only scattered zero or once during its passage through the foil. In the case of the copper foil, however, the typical electron is scattered at least once or twice, and the beam profile becomes both broader and less well defined as the occasional electron is scattered through a large enough angle to travel almost horizontally through the foil. In the gold foil, there are almost no unscattered electrons and, in fact, the scattering is now so high that a few electrons are actually seen to be backscattered from the foil. Thus while 1000 Å of carbon at 100 keV represents a specimen that is truly thin, the same thickness of gold at the same energy is beginning to take on the characteristics of a bulk (non-electron-transparent) sample. These changes clearly have important implications for electron microscope techniques such as x-ray microanalysis, in which the spatial resolution depends on the magnitude of the beam scattering. In a later chapter, this simulation will be extended to give a quantitative model of the magnitude of these effects as a function of the experimental conditions used. Experimenting with the program by trying the effect of various choices of material, foil thickness, and beam energies is a good way to start to build up a feel for the way in which electrons interact with a solid.

When the sample is thin and the beam energy is high, as in the situation discussed above, then each trajectory consists of only a small number of scattering events. Consequently the program runs rapidly and many trajectories can be computed in a relatively short time. On a 386-class MS-DOS machine fitted with a math coprocessor, it should be possible to compute the 250 trajectories for any of the cases given in no more than 30 sec. However, if in response to the question `"Is the sample bulk (y/n)?"` we choose the option `"y,"` then the situation becomes very different. Each trajectory will now continue until the electron is either backscattered or falls below the minimum energy `e_min,` which in the program above is 0.5 keV. Since the electron is typically losing energy at the rate of about 1 eV per angstrom (see Fig. 3.4), and since the mean free path for elastic scattering events is a few hundred angstroms (from Table 3.1), this suggests that even for a 20- or 30-keV incident electron, it will be necessary to compute several hundred scattering events to complete one trajectory. This is indeed the case, and it may take as long as 1 min to compute a single trajectory under the worst conditions.

Figure 3.8 shows 250 trajectories plotted for bulk samples of carbon, copper, silver, and gold with incident beam energies of 20 keV. On an IBM PS 2/70, each plot took from 2 to 5 min to calculate. The shape of the interaction volumes in the solids is seen to vary, with the choice of material, from a "teardrop" hanging from the surface for the carbon to a "squashed egg" shape pressed against the incident surface for the gold. While the size of this volume will vary with the chosen beam energy, the shape stays about the same, so that the interaction volume at 2 keV is simply a scaled down version of the equivalent volume at 20 keV. Observations of

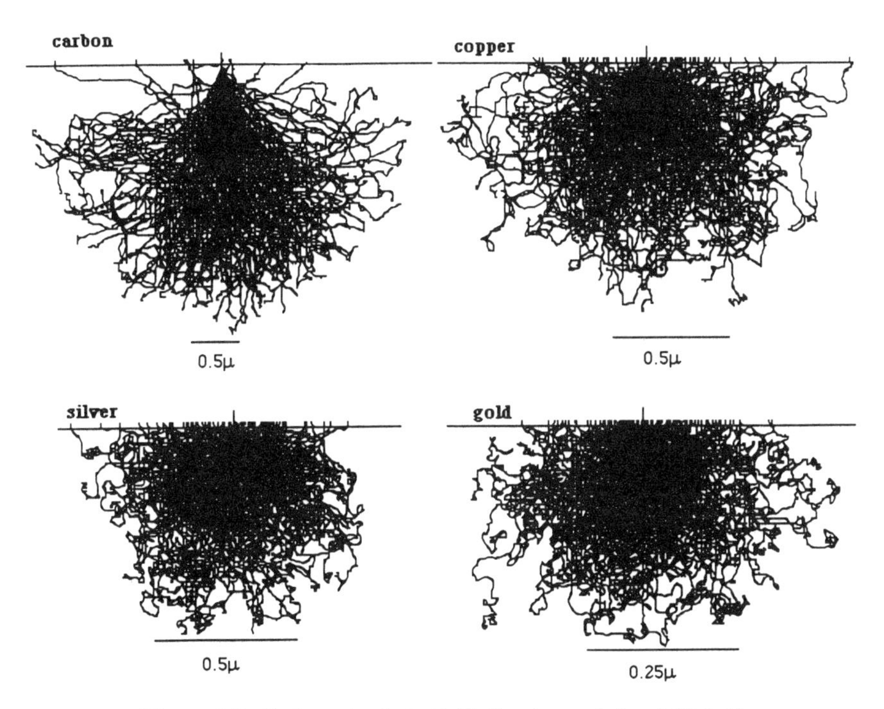

Figure 3.8. Trajectories in bulk C, Cu, Ag, and Au at 20 keV.

the trajectories while the program is running show that a disproportionately large fraction of the time is devoted to computing the last portion of each trajectory. This is because, as the energy falls to just 1 or 2 keV, the mean free path becomes short, typically 50 Å or less, and the step length [Eq. (3.7)] is only a few tens of angstroms. Although the rate of energy loss [Eq. (3.21)] is now higher than it was at the original incident energy, it has not risen as quickly as the mean free path has fallen, and hence the average energy loss per step is only a few tens of electron volts. Consequently, each successive trajectory step moves the electron by only a few angstroms and decreases its energy by a few electron volts, and a large number of computations is needed to make a relatively insignificant addition to the overall trajectory. Since the perceived utility of a Monte Carlo model is at least somewhat dependent on how fast it can produce the required information, this leisurely level of performance makes the single scattering model rather unsatisfactory in many cases. One possible solution is to choose a higher cutoff energy `e_min`, so as to eliminate many of these calculations. However—although when the incident energy is sufficiently high (say 10 keV and above) `e_min` could be set to 1 keV or more with little loss of accuracy—when the initial energy is only 2 or 3 keV, terminating the trajectory calculation at an energy only 50% below the incident value would represent a rather drastic oversimplification and a value of 0.1 keV for `e_min` would seem more appropriate.

One practical solution to this dilemma is to use a model originally described by

Archard (1961). In this we allow an electron that has reached the cutoff energy to diffuse for a total distance equal to its estimated range [obtained from Eq. (3.22)] at an energy `e_min`. This is done in two stages. First, it is allowed to continue to travel in the same direction (as defined by the direction cosines *cx, cy, cz*) in which it had been traveling when it left the Monte Carlo loop for a distance equal to

$$\text{Distance} = \frac{40}{7 * \text{atomic number}} * \text{estimated range} \tag{3.23}$$

Setting the step distance equal to this value, the new coordinates can now be found as usual, using the procedure `new_coord`. The new z coordinate `zn` is then tested to see if the electron has backscattered. If it has, then it is counted and a new trajectory started. Otherwise we set reset the coordinates as usual and then allow the electron to scatter through randomly chosen angles ϕ and ψ, where

```
cp:=2.0*RND − 1.0;        {cosine of scattering angle φ}
sp:=sqrt(1 − cp*cp);      {sine of scattering angle φ}

ga:=two_pi*RND;           {azimuthal angle ψ}
```

The distance traveled in the final segment is then a fraction (1 − (40/7Z)) of the original estimated range. Using this step length and the scattering angles given above, the new coordinates are then found, using `new_coord`, and the electron position is again tested to see if it has backscattered. If it has, then it is counted and a new trajectory is started; if not, then the trajectory is terminated and a new one is begun. This modification can be implemented by adding one function and procedure to the program given above and then calling the procedure at the appropriate point:

```
Function archard_range:extended;
   {estimates Archard diffusion range at e_min in microns}

  begin
     archard_range:=0.07*power(e_min,1.66)/density;
  end;

Procedure end_of_range;
   {apply the Archard model for electrons with energies below e_min}

  begin
   if s_en<=e_min then
   begin {travel (40/7*at_num)*archard_range in same direction}
                 step:=Archard_range*40/(7*at_num);
                 new_coord(step);
          if zn<=0 then {this one is backscattered}
                   back_scatter;
```

```
{now let it scatter isotropically and travel residual fraction of
                              range}
                              reset_next_step; {reset variables and plot it}
                               cp:=2.0*RND-1.0;
                                sp:=sqrt(1-cp*cp);
                               ga:=two_pi*RND;
                              step:=(1-(40.0/(7.0*at_num)))*archard_range;
                             new_coord(step);
                      if zn<=0 then {this one is backscattered}
                             back_scatter;
                             reset_next_step; {to plot final segment}
     end; {of the Archard diffusion model step}
  end;
```

The function computes the estimated range and the procedure then carries out the sequence of steps discussed above. The procedure `end_of_range` is called just after the end of the loop `repeat . . . until s_en<e_min`, so that it is only entered by electrons below this energy cut-off:

```
          reset_next_step; {otherwise perform another trajectory step}
    until s_en<=e_min;      {end of the repeat until loop}
      end_of_range;    {optional Archard diffusion step added here}
          num:=num+1;       {then add one to the trajectory total}
                     etc. . . .
```

The effect of this addition to the program is not dramatic, but it does provide a more physically realistic model of the beam interaction in the important low beam energy regime. For beam energies greater than a few kilo-electron volts, the diffusion model adds little to the result and can safely be dispensed with.

4
THE PLURAL SCATTERING MODEL

4.1 Introduction

We will develop in this chapter a second Monte Carlo simulation of the interactions of an electron with a solid. This new model is based on the same physical principles as the single scattering approach discussed above but makes certain simplifying assumptions that greatly reduce the computation time required. Thus, whereas the single scattering model discussed in Chap. 3 attempts to take into account every elastic scattering event encountered by the incident electron, the "plural scattering" model described here tries only to compute the net effect produced by a number of successive scattering events. In addition, we average and precalculate other parameters of the trajectory with the object of minimizing the amount of computation required during the simulation. Although these simplifications have some effect on the resultant accuracy of the model, the errors are, when the model is correctly applied, insignificant, while the saving of time is considerable. This model is, therefore, widely used, especially for problems where the incident beam is traveling into a bulk (i.e., not thin and so not electron-transparent) specimen.

4.2 Assumptions of the plural scattering model

We saw at the end of Chap. 3 that the single scattering model, while certainly appropriate to the task of modeling electron interactions in a bulk solid, is rather slow because a great number of steps in the trajectory are required to advance the electron a small distance as its energy decreases. Consequently, in applications where the majority of electrons deposit all of their energy in the target, a single scattering Monte Carlo simulation is, for many users, too slow to be either enjoyable or very useful. The procedure that we describe here follows the outline of a method that was first described by Curgenven and Duncumb (1971), and which was specifically designed to produce good data in short computing times on small computers.

The first assumption that distinguishes this model from the previous one is that we make every electron travel exactly the same total path length within the specimen before coming to rest. This distance is found by numerically evaluating the integral:

$$R_B = \int_0^E \left[\frac{-1}{\frac{dE}{dS}} \right] dE \tag{4.1}$$

which computes the total distance measured along the trajectory that is required for an electron starting with energy E to give up all of its energy. Here (dE/ds) is the stopping power given by the modified Bethe equation [Eq. (3.21)]; therefore the total distance R_B is often called, in the literature, the Bethe range, or the csda-range (continuous slowing down approximation range). Because scattering is a statistically random process, it is clear that this is only an averaged value for a given beam energy and material, but R_B is nevertheless a convenient parameter with which to characterize the general scale of the electron interaction with the specimen. In practice, it is not possible to perform the integration from zero energy as required by Eq. (4.1) because even the modified form of the Bethe equation is not valid below about 50 eV. The procedure `stop_pwr` that calculates (dE/dS) therefore checks to see whether or not the energy is below this value. If it is, then the stopping power value calculated is that for a 50 eV electron rather than the actual energy involved. This results in a small underestimation of the range (since, as shown in Chap. 3, the stopping power falls for electrons of very low energy), but the error for incident energies above a few hundred volts is negligible. The integration is carried out by the procedure `range`. Figure 4.1 plots the variation of R_B as a function of incident energy for some common materials, and it can be seen that at 10 keV, the range is

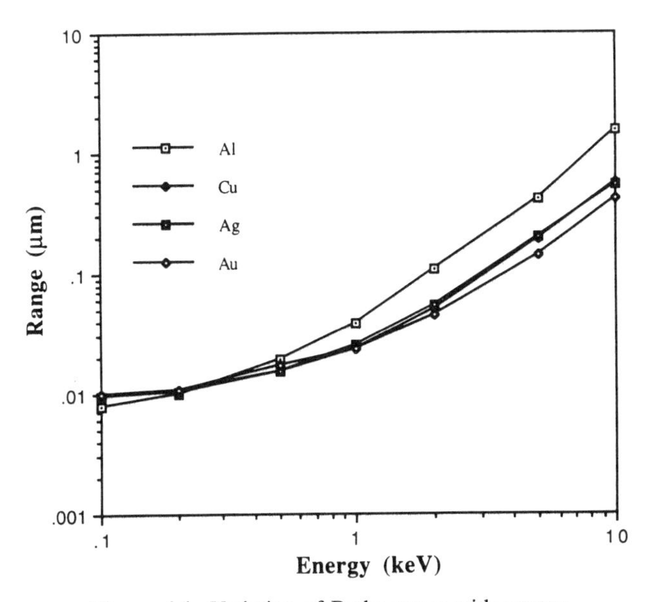

Figure 4.1. Variation of Bethe range with energy.

typically several micrometers and varies by a factor of 5 or so between the a low-atomic-number low-density element such as aluminum and a high-atomic-number high-density element such as gold. As the beam energy falls, however, the Bethe ranges converge in value and all become essentially the same, about 100 Å, for energies below a few hundred electron volts. This is because, for the lowest energies, the magnitude of the stopping power is almost independent of the choice of material.

The Bethe range is now divided into steps. In the code discussed here, 50 steps of equal length will be used. This is a compromise between the improvement in accuracy that results from a large (e.g., 100) number of steps (Bishop, 1974; Love et al., 1977), and the increase in speed to be gained from a small (e.g., 20) number of steps. An alternative approach is to limit the number of steps but vary their length with the energy of the electron (Myklebust et al., 1976), so that for high energies the step length is of the order of the instantaneous elastic mean free path but for lower energies the step length is relatively large. The total length of all the steps is, however, always kept equal to the Bethe range. In either case the effect of this procedure is to place a fixed upper limit on the number of computations required to complete the trajectory and avoid the problem, discussed in Chap. 3, of dealing with the incident electron when its energy becomes so low that the mean free path is only a few nanometers and the distance traveled in any scattering interval is an insignificant fraction of the total range.

Computer time is also saved by precalculating the energy $E[n]$ of the electron at the start of each of the n steps. Since the length of each step is known in advance, $E[n]$ can be found by numerically solving the equation

$$E[n] = E[n-1] - \int_{\text{step}} \left(\frac{dE}{dS}\right) dS \tag{4.2}$$

where (dE/dS) is again obtained from the modified Bethe equation [Eq. (3.21)]. $E[1]$ is defined as the incident electron energy E_0, so a solution of Eq. (4.2) will give $E[2]$, which, in turn, can be used to find $E[3]$ and so on until, at the end of the 50th step, we set $E[51] = 0$. This calculation is carried out by the procedure `profile`.

The electron scattering is once again described by the screened Rutherford cross section, but it is formulated in a different way. Using the notation of Chap. 3, as shown in Fig. 4.2, the scattering angle can be written as:

$$\cot\left(\frac{\phi}{2}\right) = \frac{2p}{b} \tag{4.3}$$

where p is the impact parameter (i.e., the projected nearest distance of closest approach of the electron to the scattering nucleus if no scattering was to occur) and b

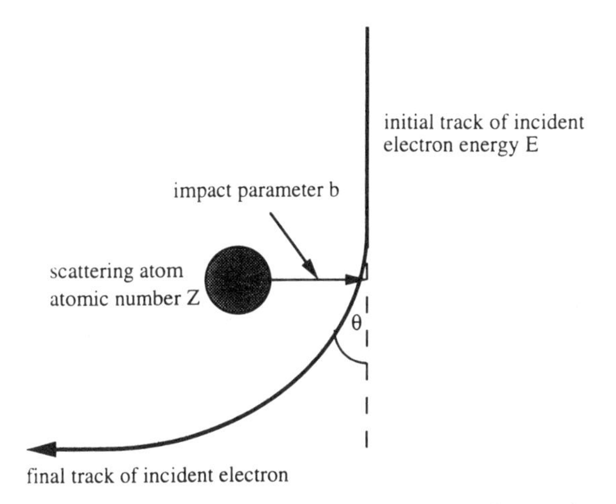

Figure 4.2. Definition of parameters used in the plural scattering model.

is 0.0144 Z/E with b in angstroms when E is in kilo-electron volts. This expression is for a single scattering event and ignores the effect of nuclear screening by orbital electrons (Curgenven and Duncumb, 1971). However, since at every energy our step length is now considerably greater than the mean free path, each electron could be scattered several times during each step, producing a net deviation that could be either larger or smaller than any of the individual deflections. This random variation is accounted for by writing

$$p = p_0\sqrt{\text{RND}} \tag{4.4}$$

where RND is the usual pseudo–random number lying in the range 0 to 1 and p_o is now the maximum impact parameter. Electrons are assumed to arrive at the nucleus placed randomly within the circle of radius p_o. The square-root function weights the distribution in such a way as to increase the probability of large p values.

The simplifications introduced into this plural scattering model, compared with those of the single scattering model of Chap. 3, are clearly fairly drastic in nature. However, the cumulative error due to these approximations can be reduced by choosing the value of p_o so that the Monte Carlo calculation gives the correct backscattering coefficient. This quantity is chosen because its value is readily available for a wide range of materials and experimental conditions (Bishop, 1966; Heinrich, 1981). Two different approaches to this have been successfully used. Love et al. (1977) rewrite Eqs. (4.3) and (4.4) in the form

$$\cot\left(\frac{\phi}{2}\right) = \cot\left(\frac{\phi_0}{2}\right)\left(\frac{E}{E_0}\right)\sqrt{(\text{RND})} \tag{4.5}$$

where, as before, E_0 is the incident beam energy and

$$\cot\left(\frac{\phi_0}{2}\right) = \frac{2p_0\,E_0}{0.0144\,Z} \tag{4.6}$$

ϕ_0 thus represents the minimum scattering angle for the incident electron with energy E_0. As can be seen from the form of the Bethe equation [Eq. (3.21)], the variation of (E/E_0) is substantially independent of the atomic number Z (the variation coming only from the mean ionization potential J, which occurs inside the logarithmic term) and the random number RND will average to a mean value of 0.5 when a large number of trials is made. It therefore follows that the backscattering coefficient η must depend only upon cot $(\phi_0/2)$. Love et al. (1977) show that cot $(\phi_0/2)$ can be written in the form

$$\cot\left(\frac{\phi_0}{2}\right) = 1\,/\,(0.02209 + 0.10716\eta + 0.03009\eta^2 + 0.37555\eta^3) \tag{4.7}$$

This method eliminates the need to know p_0 provided that the value of η for the target material being modeled is available. In particular, at low (<10 keV) beam energies, where the value of η may be varying with energy (see, for example, Reimer and Tolkamp, 1980), this approach has the benefit that realistic data may be generated without the need to use more accurate scattering cross sections.

The Love et al. approach has distinct advantages and will be considered in detail in Chap. 6. However, for this initial study we follow the suggestion by Duncumb (1977) and write

$$p_0 = 0.394\,\frac{Z^{0.4}}{E_0} \tag{4.8}$$

This expression, when used in the Monte Carlo simulation, leads to a reasonable prediction of the variation of backscattering coefficient with atomic number for both elements and [using Eq. (3.19)] compounds. [A similar expression is given by Myklebust et al. (1976); but note, however, that their paper contains a typographical error and that E_0 should appear in the numerator rather than the denominator of their variable F_i.] While this approach essentially treats p_0 as a fittable parameter to yield the correct backscattering coefficient, it has the advantage that no prior knowledge of the value of η is required.

Either of these two approaches enables the Monte Carlo simulation to backscatter the correct fraction of incident electrons. However, a closer inspection of the predictions of these calculations (Myklebust et al., 1976) shows that while the backscattering yield is correct, the energy distribution of these electrons is not in good agreement with experimental data. This is because the original form of the scattering equations [Eqs. (4.3) and (4.4)] i.e.,

$$\cot\left(\frac{\phi}{2}\right) = \frac{2\,p}{b}\sqrt{\text{RND}} \tag{4.9}$$

does not allow for a sufficient amount of small-angle scattering. Thus the largest possible value of $\cot(\phi/2)$ from Eq. (4.9) is $2\,p/b$, which, for silicon at 15 keV, corresponds to a minimum scattering angle of about 3°. All electrons will be scattered by at least this angle, and 95% of all electrons will be scattered through 5° or more. For materials with higher atomic numbers, the minimum angle will be greater and the average scattering angle higher still. In order to allow for at least some fraction of the electrons to be scattered through very small angles, we can adopt the suggestion of Myklebust et al. (1976) and write

$$\tan\left(\frac{\phi}{2}\right) = \frac{b}{2\,p}\left(\frac{1}{\sqrt{\text{RND}}} - 1\right) \tag{4.10}$$

so that, as RND tends to unity, the scattering angle can fall to zero and a fraction of the scattering events will be of the order of a few degrees or less.

As in the previous chapter, the scattered electron can travel to anywhere on the base of the cone defined by the angle ϕ, so we must also choose an azimuthal scattering angle ψ, which will be given by [Eq. (3.11)] as

$$\psi = 2\,\pi.\,\text{RND} \tag{4.11}$$

The flow of the program then closely follows that for the single scattering model, but with the difference that each trajectory is restricted to 50 steps or less:

for n **= 1 to 50**
  begin
    Get the starting energy $E[n]$ of the electron
    Get the starting coordinates x,y,z for the nth step
    Get the direction cosines cx,cy,cz relative to the initial axes
    Find the scattering angles ϕ,ψ from Eqs. (4.10) and (4.11)
    Compute final coordinates xn,yn,zn from Eqs. (3.12) to (3.15)
    Check if the electron has been backscattered
      if **yes,** exit the loop and add 1 to backscatter total
    otherwise
      Reset coordinates $x = xn$, $y = yn$, $z = zn$
      Reset direction cosines $cx = ca$, $cy = cb$, $cz = cc$
  end

The PASCAL code that implements this sequence is set out below, again using the conventions established in Chap. 2.

4.3 The plural scattering Monte Carlo code

```
Program PluralScatter;

   {this programs performs a Monte Carlo trajectory simulation using
   a screened Rutherford cross section and a plural scattering ap-
   proximation}

   {$N+}   {turn on numeric coprocessor}
   {$E+}   {install emulator package}

uses CRT,DOS,GRAPH;    {resources required}

label   exit;   {target address for goto jumps}

const   two_pi=6.28318;    {2 constant}

var
   at_num,at_wht,density,inc_energy,mn_ion_pot:extended;
   an,an_m,an_n,bk_sct,cp,c_tilt,ga nu,tilt,sp,s_tilt:extended;
   ca,cb,cc,cx,cy,cz,v1,v2,v3,v4,x,y,z,xn,yn,zn:extended;
   h_scale,m_t_step,rf,step,v_scale:extended;
   E:array [1 . . 51] of real;
   bottom,b_point,center,k,num,traj_num,top:integer;
   GraphDriver:Integer;
   GRAPHMODE:Integer;
   ErrorCode:Integer;
   Xasp,Yasp:word;
   s    :string;

Function power(mantissa,exponent:real):real;
   {because PASCAL has no exponentiation function}
  begin
   power:=exp(ln(mantissa)*exponent);
  end;

Function stop_pwr(energy:real):extended;
   {calculates the stopping power using the modified Bethe expression
     of Eq. (3.21) in units of keV/g/cm2}

var temp:real;

  begin
```

```
   if energy<0.5 then {to avoid problems as energy approaches zero}
                     energy:=0.05;
       temp:=ln(1.166*(energy+0.85*mn_ion_pot)/mn_ion_pot);

        stop_pwr:=temp*78500*at_num/(at_wht*energy);

 end;

Procedure set_up_screen;
   {gets initialization for random number generator and the input
     data required to run the program}

 begin
     ClrScr;    {Erases any data on screen display}

     GoToXY(25,1);
         writeln('Plural scattering Monte Carlo simulation');

     GoToXY(1,5);
      Write('Input beam energy in keV');
       Readln(Inc_Energy);

     GoToXY(1,7);
       Write('Target atomic number is');
         Readln(at_num);

    GoToXY(1,9);
     Write('Target atomic weight is');
      Readln(At_wht);

       {compute the mean ionization potential J using the Berger-
        Selzer analytic fit in units of keV}
                mn_ion_pot:=(9.76*at_num + (58.5/power(at
                _num'0.19)))*0.001;

    GoToXY(1,11);
     Write('Target density in g/cc is');
      Readln(Density);

       {get the beam tilt data}
            GoToXY(40,5);
            Write('Tilt angle in degrees');
            readln(tilt);
               s_tilt:=sin(tilt/57.4); {convert degrees to radians}
               c_tilt:=cos(tilt/57.4);

   {get the number of trajectories to be run in this simulation}
```

```
      GoToXY(40,7);
       write('Number of trajectories required');
        readln(traj_num);

  end;

Procedure Rutherford_Factor;

   {computes the Rutherford scattering factor for this incident ener-
   gy}

var p0:extended;

  begin

       p0:=0.394*power(at_num'0.4)*/inc_energy; {Duncumb's function}
       rf:=0.0072*at_num/(p0); {scattering constant b/2 in Eq. (4.3)}

  end;

Procedure range;
{this calculates the range in microns assuming the modified Bethe
 stopping power, Eq. (3.21)}

var energy,f,fs,Bethe_range:extended;
              l,m:integer;

  begin
        fs:=0.;          {initialize variable to be sure}

     for m:=1 to 21 do   {a Simpson's rule integration}

       begin

         energy:=(m-1)*inc_energy/20; {20 equal steps}

           f:=1/stop_pwr(energy);

         l:=2;
           if m mod 2=0 then l:=4;
            if m=1 then l:=1;
           if m=21 then l:=1;

            fs:=fs+l*f;
  end;

   {now use this to find the range and step length for these con-
   ditons}
```

```
  Bethe_range:=fs*inc_energy/60.0;              {in g/cm2}
     m_t_step:=Bethe_range/50.0;

  Bethe_range:=Bethe_range*10000.0/density;     {in microns}

  {and display this for informational purposes}

            GoToXY(40,11);
               writeln('. . . . .Range is 'Bethe_range:4:2, 'mi-
                 crons');

                 step:=Bethe_range/50.0;

 end;


Procedure profile;
   {compute 50-step energy profile for electron beam}


var A1,A2,A3,A4:extended;
              m:integer;
 begin
              E[1]:=inc_energy;

        for m:=2 to 51 do

     begin
          A1:=m_t_step*stop_pwr(E[m-1]);

               A2:=m_t_step*stop_pwr(E[m-1]-A1/2);

               A3:=m_t_step*stop_pwr(E[m-1]-A2/2);

          A4:=m_t_step*stop_pwr(e[m-1]-A3);

         E[m]:=E[m-1] - (A1 +2*A2 +2*A3 +A4)/6.;

     end;

       E[51]:=0.; {ensure electron finishes with zero energy}

          for m:=2 to 50 do {a little smoothing of the profile}
            begin
              e[m]:=(E[m] + E[m+1])/2.;
            end;

 end;
```

```
Procedure initialize;
   {sets up the graphics drivers for V5.0 TURBO PASCAL}

var
   InGraphicsMode:Boolean;
   PathToDriver:String;

 begin
    DirectVideo:=False;
     PathToDriver:='';
      GraphDriver:=detect;
       InitGraph(GraphDriver,GRAPHMODE,PathToDriver);
        SetViewPort(0,0,GetMaxX,GetMaxY,True); {clips the display}
      center:=trunc(GetMaxX/2);
     top:=trunc(GetMaxY*0.1);        {adjust to suit your screen}
    bottom:=trunc(GetMaxY*0.75);    {ditto}
 end;

Procedure set_up_graphics;

   {sets up the plotting scales, surfaces, etc., for display using
     the size of the screen as found by the initialize routine
     to scale the display properly}

var   a,b,c,d:integer;
 begin
                    a:=GetMaxX-20;    {adjust to suit your screen}
                    b:=20;            {ditto}

          Line(b,top,a,top); {plot in top surface}

           {now plot in the incident beam allowing for tilt}
           b_point:=center - trunc(38*s_tilt/c_tilt);

           Line(b_point,1,center,top); {plot beam}

              {find the aspect ratio of this display}
                    GetAspectRatio(Xasp,Yasp);
                 {set up the plotting scale}
              h_scale:=GetMaxX/(100.*step); {in pixels per micron}
                 v_scale:=h_scale*(Xasp/Yasp);

                 c:=GetMaxY-43;    {adjust for your screen}
                  d:=10;           {ditto}

      {set up to draw the trajectories completed thermometer}
      OutTextXY(trunc(center-80),bottom+15,'0% . . . . . . . 50% . . . . .
      . 100%');
```

```
      OutTextXY(trunc(center-78),bottom+28,'Trajectories
       completed');

       If h_scale<=300. then    {draw a bar 1 micron long}
                begin
       Line(d,c,d+trunc(h_scale),c); {micron bar);
          OutTextXY(d+trunc(h_scale/3),d+c,'1 μm'); {label the bar}
                  end
                     else    {draw a bar 0.1 microns long}
                  begin
                     Line(d,c,d+trunc(0.1*h_scale),c);
                     OutTextXY(d+trunc(0.03*h_scale),d+c,'0.10μm');
                  end;

  end;

Procedure xyplot(a,b,c,d:real);
   {this displays the trajectories on graphics screen}

var iy,iz,iyn,izn:integer;
  begin
     iy:=center+trunc(a*h_scale);      {plotting coordinate #1}

     iz:=top + trunc(b*v_scale);       {plotting coordinate #2}

     iyn:=center+trunc(c*h_scale);     {plotting coordinate #3}

     izn:=top + trunc(d*v_scale);      {plotting coordinate #4}

       if d=99 then izn:=top-2;        {BS plotting limit}

     {now plot this vector on the screen}

       line(iy,iz,iyn,izn);
  end;

Procedure init_counters;
   {initialize each counter since PASCAL does not do this}

  begin
         bk_sct:=0;
            num:=0;
  end;

Procedure reset_coordinates;
   {reinitialize all the electron variables at start of each new tra-
   jectory}
```

```
begin
            x:=0;
             y:=0;
             z:=0;
             cx:=0;
            cy:=s_tilt;
         cz:=c_tilt;

end;

Procedure p_scatter;
   {calculates scattering angles using the plural scattering model
   with the small angle correction as given in Eq. (4.10)}

  begin
       {first call the random number generator function}
                         nu:=sqrt(RANDOM);
                          nu:=((1/nu)-1.0);
                         an:=nu*rf/E[k];

      {and use this to find the scattering angles}
                         sp:=(an+an)/(1+(an*an));
                         cp:=(1-(an*an))/(1+(an*an));

      {and the azimuthal scattering angle}
                         ga:=two_pi*random;

  end;

Procedure new_coord(step:real);
   {gets xn,yn,zn from x,y,z and scattering angles using Eqs. (3.12)
   to 3.15)}

var an_n,an_m,v1,v2,v3,v4:extended;
  begin
    {the coordinate rotation angles are}
                    if cz=0 then cz:=0.000001; {avoid division by
                    zero}
                         an_m:=(-cx/cz);
                         an_n:=1.0/sqrt(1+(an_m*an_m));

    {save computation time by getting all the transcendentals first}
                         v1:=an_n*sp;
                          v2:=an_m*an_n*sp;
                          v3:=cos(ga);
                         v4:=sin(ga);
```

```
   {find the new direction cosines}
                  ca:=(cx*cp) + (v1*v3) + (cy*v2*v4);
                   cb:=(cy*cp) + (v4*(cz*v1 - cx*v2));
                  cc:=(cz*cp) + (v2*v3) - (cy*v1*v4);

   {and get the new coordinates}
                  xn:=x + step*ca;
                   yn:=y + step*cb;
                  zn:=z + step*cc;
  end;

Procedure reset_next_step;
   {rests variables for next trajectory step}

  begin
                  xyplot(y,z,yn,zn); {plots this step}
                      cx:=ca;
                       cy:=cb;
                        cz:=cc;
                        x:=xn;
                       y:=yn;
                      z:=zn;

  end;

Procedure back_scatter;
   {handles special case of a backscattered electron}
  begin
          bk_sct:=bk_sct+1;    {add one to counter}
          num:=num+1;          {add one to total}
          xyplot(y,z,yn,99);   {plot BS exit}
  end;

Procedure show_traj_num;
   {update the thermometer display}

var a,b:integer;
  begin
     a:=trunc(center-80);    {to match position of thermometer}
     b:=bottom+23;           {ditto}

     line(a,b,a+trunc(165*(num/traj_num)),b);    {draw it in}

  end;

Procedure display_backscattering;
   {draws a thermometer to display the backscattering coefficient}
```

```
label   hang;
var   a,b:integer;

  begin
            a:=GetMaxX-180;        {adjust to suit your screen}
            b:=bottom+23;          {ditto}

            OutTextXY(a,b-8''0 . . . . 0.25 . . . 0.5 . . . 0.75');
             OutTextXY(a+18,b+5 'BS coefficient');
            Line(a,b,trunc(a+(bk_sct/traj_num)*220),b); {draw it}
      hang:
           if (not keypressed) then goto hang; {freeze display on
           screen}
          CloseGraph; {shut down the graphics unit}
  end;

{
*************************************************
*      this is the start of the main program      *
*************************************************

}
begin
            set_up_screen;         {get input data}

            Rutherford_factor;     {and screening factor}

   {now get the range and step length in microns for these conditons}

            range;
            profile;

            initialize;          {initialize the graphics screen}
            set_up_graphics;     {and draw on it}

            randomize;           {reseed random number generator}

{************************************************
 *              the Monte Carlo loop              *
 ************************************************}

                 init_counters; {reset counters}

            while num < traj_num do
  begin
```

```
                    reset_coordinates;

                   for k:=1 to 50 do

      begin
                      p_scatter;          {find the scattering angles}

                      new_coord(step);    {find where electron goes}

   {program specific code will go here}

   {now test for the electro nposition in the sample}
          if zn<=0, then    {this one is backscattered}
            begin
             back_scatter;
             gotoexit;
            end
          else              {it is still in the target}
            reset_next_step;

      end;          {of the 50-step loop}

          num:=num+1;       {add one to trajectory total}

        exit:               {end of goto jumps}

        show_traj_num       {update the trajectory number display}

  end;     {of the Monte Carlo loop}

{**************************************************
 *            end of the Monte Carlo loop        *
 **************************************************}
        display_backscattering; {show the computed BS coefficient}

end.
```

4.4 Notes on the procedures and functions used in the program

The program starts, as in the case of the single scattering model, with the pragmas {\$*N*+} and {\$*E*+}, which select the use of the math coprocessor and the emulator package. The resources, labels, constants, and variables follow. The first two functions in `PluralScatter`, `POWER`, which provides an exponentiation capability for PASCAL, and STOP_PWR, which calculates the electron stopping power in units of keV/g/cm^2 using Eq. (3.21), are identical to those in SingleScatter, since, wherever possible, it is advantageous to reuse program code. Next, set_up _screen, allows for the input of the various pieces of data to describe the experimental conditions to be modeled. In addition to the usual data about the specimen (atomic number, atomic weight, and density) and the beam energy, the option is given to change the angle of incidence of the beam to the specimen. The variable Tilt, which describes this angle, is given in degrees, so the procedure must first convert it to radians. Using the convention for the Cartesian axes described in Chap. 3, the three direction cosines for the incident beam will then be CX = 0, CY = sin(Tilt), CZ = cos(Tilt). The procedure calculates and stores these values.

The procedure Rutherford_Factor calculates the energy-independent quantities that appear in the scattering-angle Eqs. (4.3), (4.4), and (4.10). The variable p0 is Duncumb's suggested fitting function to the backscatter data as given in Eq. (4.8); rf then represents the quantity $(b/2p_0)$, where b is 0.0144 Z, which is used in Eqs. (4.3) and (4.10). The range of the electron is now calculated in the procedure range using Eq. (4.1). We first divide the total energy range of the electron, from its starting value inc_energy to zero, into 20 equal steps. At each of these 21 energy values, the quantity $1/(dE/dS)$ is then found from Eq. (3.21), calculated using the function stop_pwr. Finally, Simpson's rule (Press et al., 1986) is used to evaluate the integral, giving a result in units of mass per unit area (g/cm^2). To obtain the Bethe range, this quantity is divided by the density of the specimen (g/cm^3) and the result is converted to micrometers and printed out on the input display screen. The step_length is then set equal to one-fiftieth of this range.

The variation of energy of the electron can now be precalculated using the procedure profile, which solves Eq. 4.2 using a Runge-Kutta method (Press et al., 1986) to obtain the electron energy at the end of each of the 50 steps into which the trajectory has been divided. Starting from the incident beam energy at the beginning of the first step, the procedure obtains four different estimates for the energy lost in traveling one step length. Since the stopping power itself varies with energy, the first estimate gets the energy loss ΔE_1 along the step, assuming the stopping power appropriate to the starting energy inc_energy. The next estimate subtracts half of the calculated energy loss ΔE_1 from the starting energy and, using this energy value (i.e., inc_energy- $\Delta E_1/2$), finds the stopping power and hence a second estimate ΔE_2 for the energy loss along the step. This value is, in turn, used to find the stopping power at the energy (inc_energy- $\Delta E_2/2$) and the energy loss ΔE_3 along the step. Finally, the stopping power at the energy (inc_energy- ΔE_3) is found and the energy loss ΔE_4 is computed. The actual energy at the end of the first step is then the starting energy inc_energy minus the weighted average of the various energy-loss estimates. This value is then used as the starting energy for the next step, and the process is repeated until the end of the 50th step, which

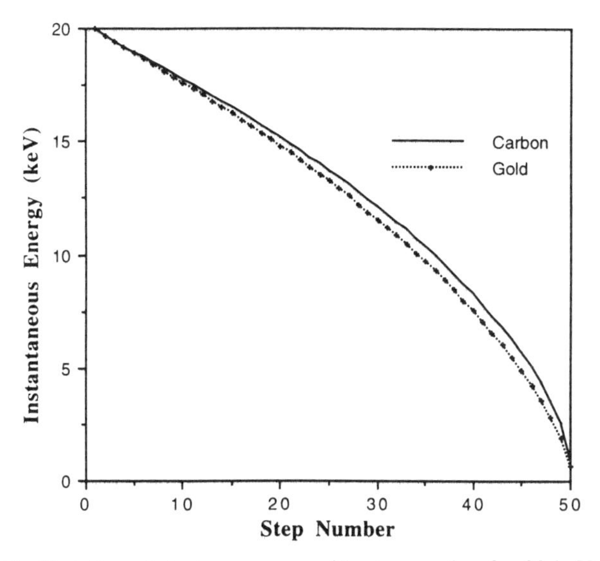

Figure 4.3. Variation of electron energy with step number for 20-keV incidence.

sets the energy equal to zero. Figure 4.3 plots $E[n]$, the energy at the start of the nth step, as a function of n for 20-keV electrons incident on carbon and gold. It can be seen that, when plotted as a function of the fraction of the range traveled, electrons in both materials lose energy at about the same rate and that the rate of energy loss is quite nonlinear (i.e., the electrons travel about two-thirds of their range before losing half of their initial energy but lose the last 25% of their energy in the last 10% of the steps).

The procedure `initialize` is identical to that used previously and again determines the plotting size of the screen and the location of `center`, the horizontal midpoint of the screen, and `top`, the location at which the entrance surface of the specimen is to be drawn. The procedure `set_up_graphics` uses these values to plot in the top surface and the incident beam, allowing for this beam to be tilted relative to the surface normal. The horizontal plotting scale `h_scale` is set equal to the width of the screen (GetMaxX) divided by twice the Bethe range (pixels per micron), and the corresponding plotting scale `v_scale` in the vertical direction is found by multiplying `h_scale` by the aspect ratio of the screen to ensure a correctly proportioned display. Finally, a micron marker and a "trajectories completed" thermometer are constructed. As before, while the code provided will correctly position these for any type of graphics screen that is used, their exact location can be varied to suit individual preference by adjusting the values of the local variables `c` and `d`.

Procedure `p_scatter` uses Eq. (4.10) to calculate $\tan(\phi/2)$, where ϕ is the scattering angle defined in Fig. 4.2. The values of $\sin(\phi)$ and $\cos(\phi)$ are then obtained by standard trigonometric formulas from the value of $\tan(\phi/2)$,

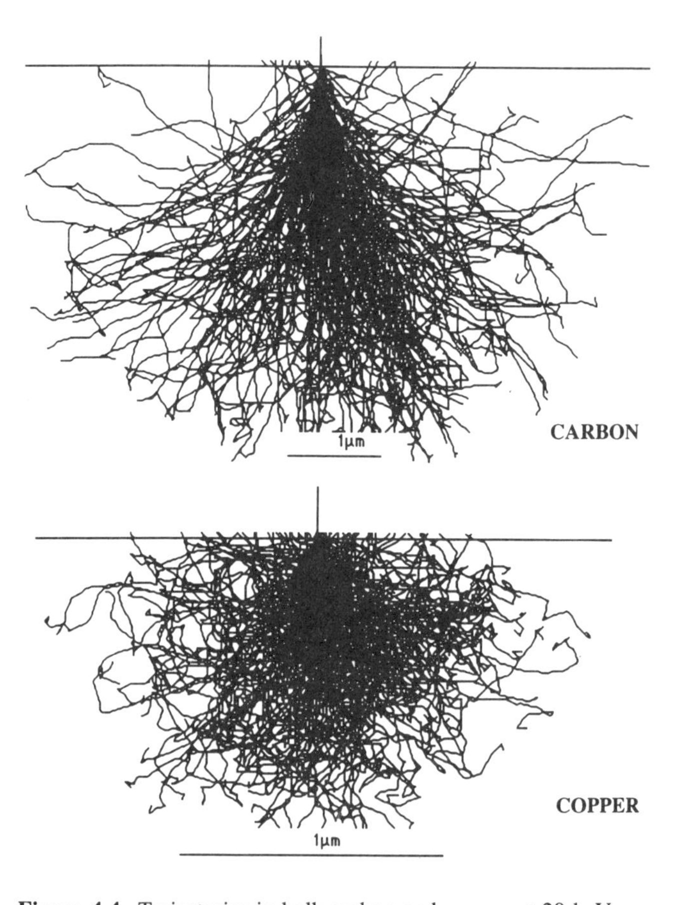

Figure 4.4. Trajectories in bulk carbon and copper at 20 keV.

and the azimuthal angle ψ is found from Eq. (3.11). These angles are then passed to the procedure `new_coord`, identical to the version in Chap. 3, which determines the coordinates of the end point of this step. The final procedures `reset_next_step`, `back_scatter`, `show_traj_num`, and `display_backscattering` handle the various bookkeeping details for each of the trajectories and update the screen display.

The main body of the program, starting at **`begin`** and finishing at **`end`** simply calls the procedures as they are required, looping until the required number of trajectories have been computed. In the center of this loop is the place, after the new position of the electron has been determined, where code to compute specific features of the electron interaction (such as x-rays or secondary electrons) can be placed. The electron is then be tested to see whether it has been backscattered and, if so, the `back_scatter` procedure is called; otherwise, the electron proceeds to the next step. After completion, the program displays the fractional yield of backscattered electrons and then terminates.

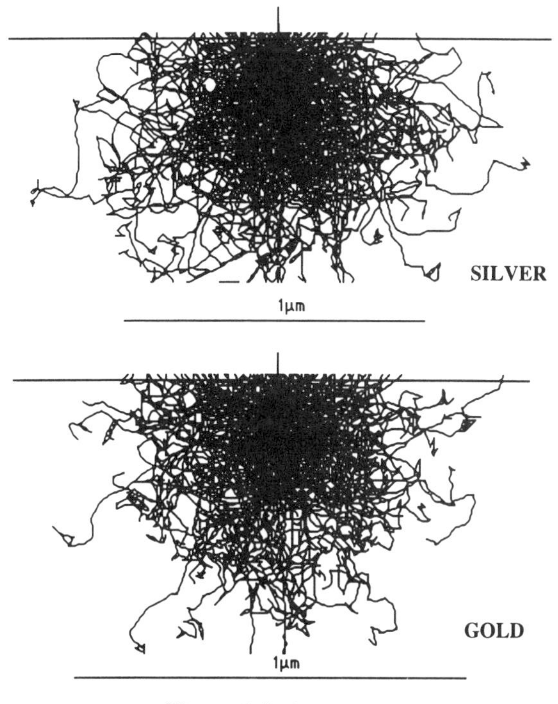

Figure 4.4. *(continued)*

4.5 Running the program

The program can be run directly from the disk if this is available by typing PC_MC at the ">" prompt and then hitting return. Once the required input data (atomic number, atomic weight, density, and beam energy as before) have been provided, the program will start running. A comparison of the speed of this program with the equivalent computation for a bulk specimen using `SingleScatter` will show that `PluralScatter` is 10 to 20 times faster, because a maximum of only 50 steps is calculated for each trajectory. Figure 4.4 shows computed trajectory plots for 20-keV electrons in carbon, copper, silver, and gold to correspond with those in Fig. 3.7. Despite the simplifying assumptions introduced by the plural scattering model, it is evident that the basic details of the interaction are well accounted for. In later chapters, we examine in more detail some verifiable predictions of both models and find that, when used with proper precautions, either can give results of good accuracy. It should be noted that one consequence of the model used here is that the shape of the interaction volume is independent of the actual incident beam energy and depends only the material of the target. The form of the interaction volume does, however, depend on the angle of incidence of the beam, as shown in Fig. 4.5, which plots trajectories for 30° and 60° angles of incidence in copper. Note that the

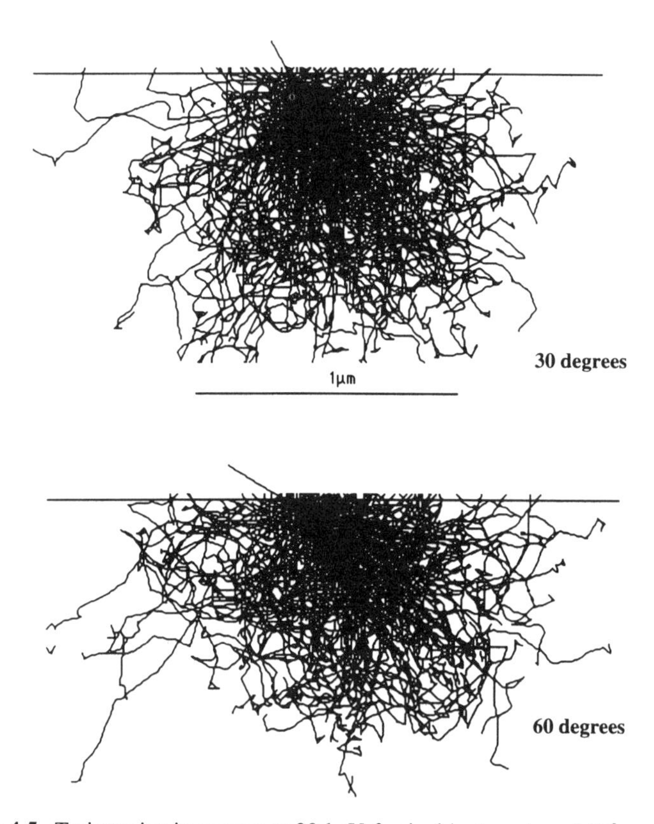

Figure 4.5. Trajectories in copper at 20 keV for incident angles of 30° and 60°.

maximum extension of the interaction follows the general direction of the incident beam, so that the volume becomes elongated in the direction of the beam. The lateral extent of the interaction is therefore quite different in the plane containing the incident beam and in the direction normal to this, and this will be evident in both imaging and microanalysis performed under these conditions.

5

THE PRACTICAL APPLICATION OF MONTE CARLO MODELS

5.1 General considerations

The remainder of this book is devoted to the application of Monte Carlo models to a variety of problems in electron microscopy and microanalysis. The programs developed in Chaps. 3 and 4 provide only the framework that models the basic details of the electron interaction. We will now develop a variety of algorithms that, when implemented as PASCAL procedures and functions, can be used to transform these generic Monte Carlo models into programs customized to solve a particular problem. Before proceeding to these applications, however, there are a few practical considerations that are worth discussing.

5.2 Which type of Monte Carlo model should be used?

We have developed two types of Monte Carlo model, the single scattering model, which attempts to account for every elastic interaction suffered by the incident electron, and the plural scattering model, which considers only the resultant effect of scattering events occurring within some specified segment of the electron trajectory. These model share much of the same physics, and so, when properly used, can be expected to give comparable results. It is, however, necessary to decide what constitutes proper usage. An important general property of these models is what we shall call, by analogy with photographic film, their *granularity.* This concept expresses the idea that the Monte Carlo model is taking what is in reality a continuous sequence of scattering events and modeling it as a discrete series of independent events, just as a piece of film takes a picture and breaks it down into fragments the size of the grains making up the emulsion. The finer the grain size of the film, the higher the resolution of the image; and the finer the granularity (i.e., the step size) of the Monte Carlo simulation, the better the quality of the model generated. In the case of the single scattering model, the step size is of the order of the elastic mean free path and thus (depending on energy) is between a few nanometers and a few tens of nanometers, while for the plural scattering model, the step size is a fraction of the Bethe range and varies from tens of nanometers to fractions of a micrometer.

In modeling some effect, it is therefore necessary to ensure that the granularity of the model is sufficient to resolve the kind of effect that is being studied. It is obvious that a plural scattering model, which at an energy of 100 keV would have a step length of close to a micron, would not be suitable for studies of electron scattering in a thin foil only a few hundred angstroms thick—the data would be excessively "grainy." But care must also be taken to see that the granularity is not too small. For example, at low beam energies, the step length in the single scattering model is only a few angstroms and so is approaching atomistic dimensions. Caution must be exercised in this condition, because a fundamental assumption of the Monte Carlo method is that the sample can be considered to be a structureless continuum (sometimes called a "jellium"), and this will only be true when the step length is much greater than either atomic or crystal lattice dimensions. Conclusions drawn about electron behavior under these extreme conditions may therefore be in error, and it would actually be better to use a more granular model. An appropriate test is to determine the critical dimensions of the phenomena or experiment to be simulated and ensure that the model chosen has sufficient granularity to resolve the effect. At the same time we must also ensure that the scale is not so far below that of the problem that excessive time is wasted in computation. For example, the modeling of backscattered signals from bulk specimens can usually be done with a plural scattering model, because the dimensions of the sample and its features or inclusions are usually large compared to the step size. But in modeling secondary electron production from the same specimen, it may be necessary to use a single scattering model, because the scale of the secondary emission (set by the escape depth of secondary electrons) is only a few nanometers. Using a single scattering model in the first case would probably give the same result, but at the expense of much extra computing time; while using a plural scattering model in the second case might give a result that is accurate on the scale of the step length (i.e., fractions of a micron) but inaccurate on the scale of a few tens of angstroms.

5.3 Customizing the generic programs

The aim of the subsequent chapters of this book is to develop a library of procedures and functions that, when coupled with one or other of the basic Monte Carlo models, will realistically and accurately simulate the physical process of interest. These procedures can either be added to the basic programs by typing them in, or the code fragments or complete programs can be taken from the disk. In most cases, the customization requires only the addition of the procedure of interest to the body of the program together with the declaration at the start of the program of any additional global variables, the addition of a call to this procedure (or procedures) at the appropriate point within the Monte Carlo loop, and the addition of any special tests (for example, determining the position of the electron relative to some feature or surface) that are require to model the specific event of interest. As a practical matter, it is usually most convenient to modify programs in a three-step process. First, add the procedures and their associated global variables to the program and check that it still compiles correctly. Next, add the code that performs any specific tests made prior to the calling of the procedure and again check that the program compiles and runs. Finally, add the call(s) to the procedure itself and see that the program functions as expected. Breaking the process into these steps greatly eases the problem of debugging the program in the event of an error, because only one change has been made to the program at each iteration.

5.4 The "all purpose" program

An apparently irresistible temptation in Monte Carlo modeling is the urge to construct an all-purpose simulation that can answer any question about any effect from a specimen of arbitrary geometry and inhomogeneity. The problem with producing such an oracle is that the more general a program must become, the more convoluted must be its logic, and the difficulty of debugging a program rises exponentially with its complexity. It is not just that a general program is longer but that it has built into it a multiplicity of options (i.e., `if . . . then . . . else. . .` statements) that can (and, perversely, quite often do) interact with one another in unexpected ways. Thus, while a program may function quite normally and correctly for one set of conditions, another and apparently equally reasonable set of conditions may cause a malfunction or, worse, a subtle error in the output data.

The most reasonable advice is to construct a new simulation for each problem that is to be solved. For example, if we are interested in the x-ray production from features in the shape of either spheres or cubes, it is best to write one program that tackles the case of a spherical sample and another separate program that considers the case of the cube. Since a high fraction of the code will be identical in both cases, the additional time taken in constructing a second program is small. But the elimination of complicated and messy tests related to the specimen shape will make the code easier to write and follow, much easier to debug, and (very likely) faster in execution because some additional shortcuts, simplifications, and optimizations may become possible. In summary, a Monte Carlo model is most efficient and effective when it is constructed as a special tool to solve a particular problem.

5.5 The applicability of Monte Carlo techniques

It has been said that "for a two-year-old child with a hammer, everything in the world is a nail." Monte Carlo users can occasionally be guilty of the same restricted vision. While the technique is of very wide and general applicability, there are problems that cannot be solved by this type of approach. Specifically, the methods discussed here cannot be applied to problems that violate the basic assumptions of the models; for example, electron channeling and most problems associated with imaging in the transmission electron microscope cannot be investigated because they involve the crystalline nature of a sample that the Monte Carlo simulation is treating as structureless. Nor can Monte Carlo methods be applied to problems that rely on effects, such as electron diffraction or the effect of electron spin polarization, which are not accounted for in the physics built into the programs. Finally, the programs may be of only limited value at very low (below say 0.5 keV) and very high (above 500 keV) beam energies because important physical effects—such as the consequences of relativistic or nuclear interactions—have been omitted from the physics. There are also occasions—such as scanning electron microscope (SEM) operation at low magnifications, where the details of the electron beam interaction are less significant than purely geometrical effects—where a Monte Carlo approach is simply not as appropriate as a simpler and faster analytical model might be. Within these boundaries, however, the scope for the use of Monte Carlo simulation techniques is wide and, as demonstrated in the subsequent chapters, the programs developed here can be used effectively to answer many questions about to solve many commonly encountered problems in microscopy and microanalysis.

6
BACKSCATTERED ELECTRONS

6.1 Backscattered electrons

In this chapter, we will concentrate on simulations associated with various aspects of backscattered electrons. We will define backscattered electrons as those incident electrons that are scattered out of the target after suffering deflections through such an angle that they leave the material on the side by which they entered (i.e., for normal incidence, the minimum scattering angle required is 90°). In practice, it is often more useful to define backscattered electrons as those incident electrons that are scattered in the target in such a way as to be collectible by a suitable detector placed on the incident beam side of the specimen. This, of course, implies that the apparent magnitude of the backscattering will be affected by the size and position of the detector relative to the specimen and incident beam. To avoid confusion, we will separate these cases by calling the total number of backscattered electrons per incident electron the backscattering yield η and the number per incident electron as measured by some specified detector the backscattered signal.

Since every Monte Carlo model tracks each incident electron in its passage through the target, the determination of whether or not a given electron is backscattered is inherent in the simulation. At the same time, the backscattering yield η is a convenient macroscopic measure of the interaction of the electron beam with the target and forms a useful experimental test of the predictions of a Monte Carlo model. Before attempting to simulate details of backscattered imaging from complex and inhomogeneous samples, we must therefore demonstrate that the models developed in the previous chapters correctly match experimental values for the backscattering yield from planar and homogeneous materials.

6.2 Testing the Monte Carlo models of backscattering

The variation of backscattering yield with atomic number was first established nearly a century ago (Starke, 1898; Campbell-Swinton, 1899). Figure 6.1 shows a compilation of some modern backscattering yield data plotted against atomic number for an incident beam energy of 10 keV using results taken from Bishop (1966), Drescher et al. (1970), Hunger and Kuchler (1979), Neubert and Rogaschewski (1980), Reimer and Tolkamp (1980), and Heinrich (1981). η is seen to vary consid-

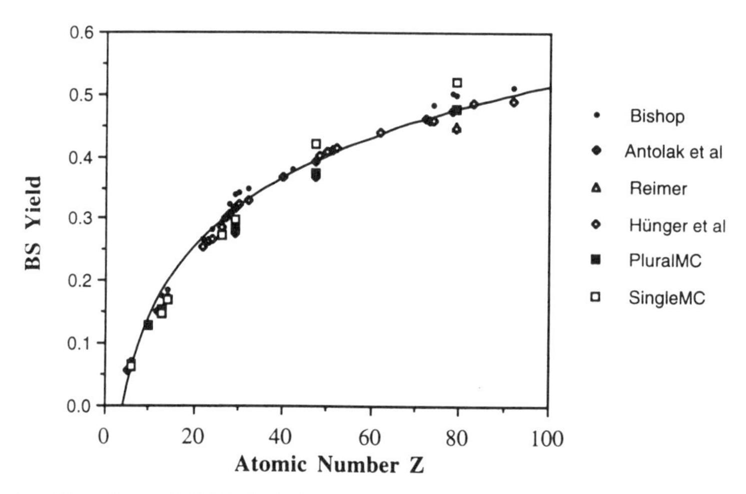

Figure 6.1. Experimental BSE yield data at 10 keV and corresponding Monte Carlo values.

erably across the periodic table, increasing from about 0.06 for carbon to about 0.5 for uranium. The variation with atomic number is generally smooth, although—as pointed out by Bishop (1966)—the slope of the curve changes discontinuously at about $Z = 30$ and again at about $Z = 60$. These changes have been associated with the binding energies of the atomic electrons. However, while the general form of the variation is evident, it is clear that there is significant scatter between different pieces of data. Although most of the cited authors claim measurement accuracies of better than 5%, variations between comparable values are as much as 20% in some cases. These discrepancies probably arise from the variety of methods employed to measure η and in particular from the success with which the effect of secondary electrons can be removed from the data. A complete set of experimental backscattered data measurements is available in Joy (1993).

The backscattering coefficient η—computed using either the single or the plural scattering Monte Carlo models (1000 trajectories per point)—is seen to be in good agreement with the experimental data, either value lying within the spread of measured values. The single scattering values, especially for $Z > 40$, tend to be slightly on the high side of the best-fit trend line through the data, but the sense and magnitude of the deviation is not systematic. The excellent agreement for either of the models is, of course, gratifying, but it is not unexpected, since both models contain a parameter that can be selected so as to match the experimental backscattering yields. In the case of the single scattering model, the screening parameter α can be adjusted to ensure a good fit to measured data. The expression for α as given in Eq. (3.2) and used in our model is that suggested by Bishop (1976). Similarly, in the plural scattering model, the minimum impact parameter p_0 [Eqs. (4.3) and (4.4)], when used in the form given in Eq. (4.8), again allows the computed backscatter data to be fitted to the experimental values. In both models, the parameter to be

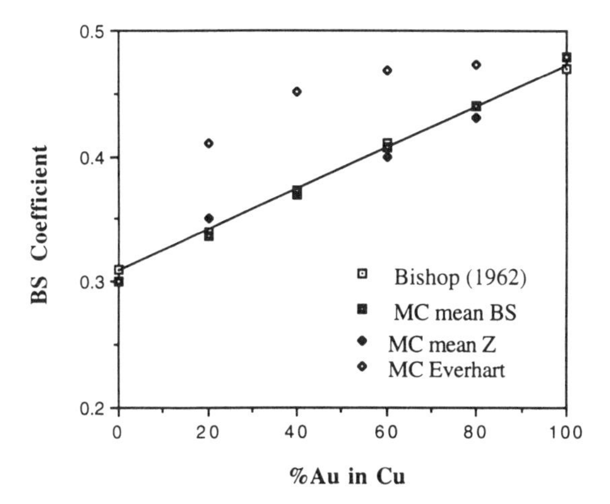

Figure 6.2. Variation of BS yield from gold/copper solid solution and corresponding Monte Carlo simulated data using various models.

adjusted is a simple function of the atomic number Z; hence improving the fit to one chosen data value will usually worsen the fit at some other value. The suggested values represent the best overall fit to the measured values at 10 keV.

It is also necessary to consider in somewhat more detail the variation of η with composition when the target is not a pure element (Herrman and Reimer, 1984.) Figure 6.2 plots the variation of the backscattering coefficient for a solid solution of gold in copper (data taken from Bishop, 1966). As the atomic percentage of gold increases from 0% to 100%, the measured backscattering coefficient increases linearly from the expected value for pure copper to the expected value for pure gold. If we have a material with the chemical formula $\Sigma_i a_i X_i$ where the X_i are elements of the periodic table with corresponding atomic weights A_i, and the a_i are their valences in the chemical formula (e.g., in H_2O, $a_1 = 2$, $a_2 = 1$, $X_1 = H$, $X_2 = O$, $A_1 = 1$, $A_2 = 16$), then we can define a mass concentration c_i from the formula $c_i = (a_i A_i / \Sigma_i a_i A_i)$, where $\Sigma_i c_i = 1$. The data shown in Fig. 6.2 represents a special case of a general result (Castaing, 1960; Heinrich, 1981), which states that the backscattering coefficient η_{mix} for a mixture of elements is given by the relation:

$$\eta_{mix} = \sum_{i=1}^{i=n} c_i \, \eta_i \tag{6.1}$$

In simple situations, it is clearly possible to find the value of η_{mix} by finding the individual values of η_i for the components, using the simulations already described and the summing them using Eq. (6.1). As Fig. 6.2 shows, this provides a very close fit to the experimental data. This procedure is, however, rather tedious if the chemistry of the target is complex (except for the case of the HKLCS model, discussed below); instead, we might try to obtain the correct value of η_{mix} by making some assumption about an effective or average atomic number for the compound of interest. The simplest suggestion (Müller, 1954) is to use the mean value of atomic number Z_{mix} for the compound—i.e.,

$$Z_{mix} = \sum_{i=1}^{i=n} c_i Z_i \tag{6.2}$$

The effective atomic weight A_{mix} is found in a similar fashion, and the value of the density used is the measured value for the compound. Figure 6.2 shows the results of employing this assumption and the plural scattering Monte Carlo model of Chap. 4 to model the copper-gold system. The agreement between the experimental and computed data is adequate to good over the full range from 0% to 100% gold, indicating that the accuracy of this simple approximation is likely to be adequate for most purposes. Alternatively, because electron stopping powers computed from the Bethe equation are additive, Everhart (1960) suggested that a more physically exact expression for Z_{mix} would be

$$Z_{mix} = \frac{\sum_{i=1}^{i=n} c_i Z_i^2}{\sum_{i=1}^{i=n} c_i Z_i} \tag{6.3}$$

However, as shown in Fig. 6.2, application of this rule to the copper-gold system produces a considerable deviation between the predicted and experimental data, suggesting that Eq. (6.2) is a more useful expression. Similar results are found in more complex systems. Figure 6.3 plots the experimentally determined backscattering coefficient for the $Al_xGa_{1-x}As$ system as a function of the mole fraction of aluminum (Sercel et al., 1989). Superimposed on this plot are the computed predictions for the backscattering, using Eqs. (6.2) and (6.3). In this example, for which the differences between the atomic numbers are relatively small and the composition range is limited, the difference between the two models is not as marked, but it is clear that the simpler expression of Eq. (6.2) is still at least as good as the more complicated expression of Eq. (6.3). In the rest of this volume, therefore, compounds will be computed either assuming the use of Eq. (6.2) or from the HKLCS procedure, discussed below.

The backscattering coefficient η is not constant with energy, as is often stated in introductory textbooks, but varies in a manner that depends on both the energy

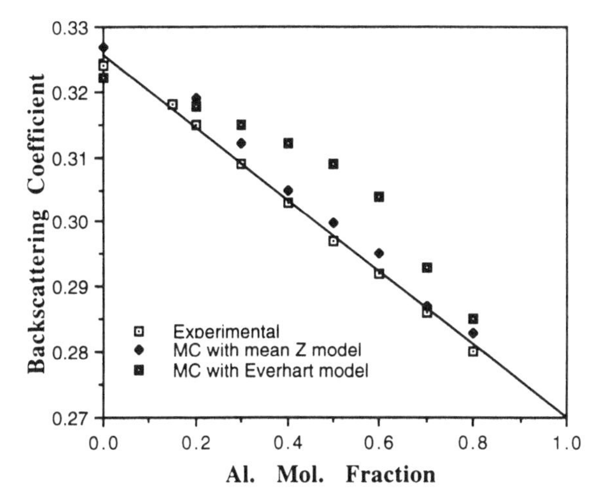

Figure 6.3. Variation of BS coefficient for Al in AlGaAs system (Sercel et al., 1989) and simulated MC data using different models.

and the atomic number. Figure 6.4 plots data from Hunger and Küchler (1979) for η as a function of Z at 4 and 40 keV. It can be seen that while the general form of the variations at the two energies is similar, the absolute values change significantly. On moving from 4 to 40 keV, the backscattering coefficient of the lightest elements falls by a factor of up to two times, while that for the heavier elements rises by as much as 25%. The plots for 4 and 40 keV actually cross at about Z = 40. The detailed form of this variation is shown more clearly in Fig. 6.5, which plots the variation of

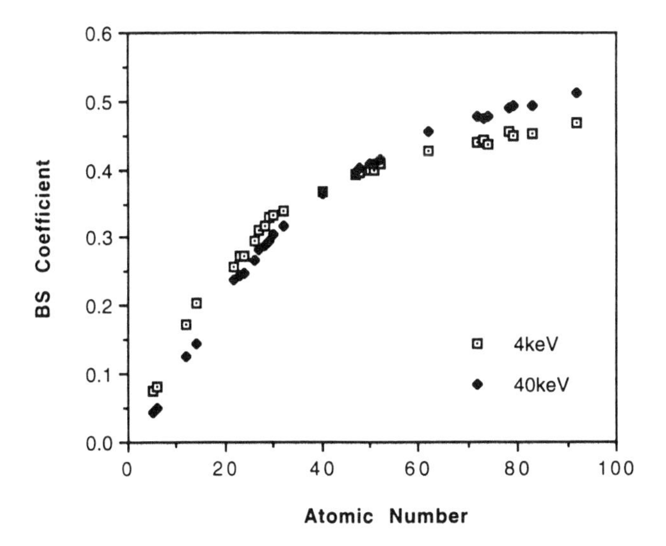

Figure 6.4. Variation of BS yield with energy.

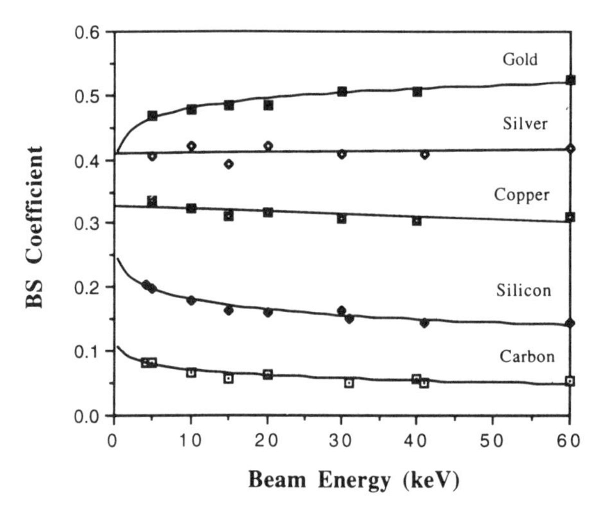

Figure 6.5. Variation of BS yield with energy.

η for carbon, silicon, copper, silver, and gold as a function of beam energy. It is clear that at the lowest beam energies (i.e., <1 keV), the bulk backscattering coefficients for all elements tend to converge to values between 0.2 and 0.4. For higher energies (>50 keV) the backscattering coefficients for light elements tend to decrease by about 10% for each factor of 10 increase in beam energy, while those for heavier elements remain about constant (Antolak and Williamson, 1985). Hunger and Küchler (1979) have derived an analytical expression that predicts, fairly accurately, values of η as a function of both the atomic number Z and the incident energy E (in kilo-electron volts):

$$\eta(Z,E) = E^{m(Z)} * C(Z) \tag{6.4}$$

where

$$m(z) = 0.1382 - \frac{0.9211}{Z^{0.5}} \tag{6.5}$$

$$C(Z) = 0.1904 - 0.2236 \ln Z + 0.1292 (\ln Z)^2 - 0.01491 (\ln Z)^3 \tag{6.6}$$

Neither of the two Monte Carlo models discussed in the previous chapters can predict this type of behavior. In fact, a plot of η versus energy using either of our models will show that, over the energy range 5 to 40 keV, the predicted backscattering coefficient is, to within the expected statistical error, independent of energy. (For lower energies, the values from the single scattering model may vary, but this is because the computation of trajectories is terminated at an arbitrary cut-off energy, which is a significant fraction of the incident beam energy). There are several reasons for this disagreement between experimental and computed data. First, the parameters in both models that adjust the backscattering values to match experimental data at some nominal energy are designed to hold the value of η constant with beam energy, since, until quite recently, low-energy electron interactions were of little interest; therefore, the assumption that η was constant with energy was a reasonable simplification. Second, as noted in discussing the data in Fig. 6.1 above, the quality of the experimental data on backscattering coefficients is rather mixed, and the magnitude of the variation of η with energy for most elements is less than the spread between different published values at the same energy. Consequently, many workers have preferred to take η as constant until more accurate experimental values were determined. Finally, and of more fundamental significance, both models use the screened Rutherford cross section as the basis for the model of electron scattering. For a wide range of conditions—that is, for elements of low and medium atomic number, electron energies greater than 10 keV, and small scattering angles—the Rutherford cross section is a good approximation; but for heavy elements, energies below 10 keV, and large scattering angles, the Rutherford cross section can be significantly in error. It must, instead, be replaced by the Mott cross section, which takes into account spin-orbit coupling in the solution of the relativistic Dirac equation (Reimer and Lodding, 1984). In general terms, the Mott cross section is essential in applications where the interaction of interest consists of a single elastic large-angle scattering event (e.g., in backscattering from thin foils or in calculations of the energy spectrum of backscattered electrons) and is desirable for all low-energy applications. But for effects caused by electron diffusion and plural scattering (e.g., backscattering from bulk samples, x-ray production, and secondary electron generation), the Rutherford cross section gives rather good agreement with experimental data except at low beam energies. The Mott cross section is considered in more detail in the final chapter of this book, but elsewhere the Rutherford model will be used because, unlike the Mott model, which requires the generation of extensive tables of data, it can be expressed analytically.

With the increased interest in low-energy scanning microscopy and microanalysis, it would, however, be desirable to try and modify the models developed earlier to correctly predict the variation of η with energy so as to obtain some of the benefits of the Mott cross section without the extra effort that is required. This is not readily possible for the single scattering model because of the necessity, discussed in Chap. 3, of terminating the electrons' trajectories at some predetermined cutoff energy, usually 0.5 keV. At low energies, this cutoff is a significant fraction of the incident energy and hence the precision of the computation is doubtful. For the plural scattering model, however, a variation of Z with E can readily be achieved. As discussed in Chap. 4, the scattering angle ϕ can be written in the form [using Eqs. (4.5) and (4.10)]:

$$\tan\left(\frac{\phi}{2}\right) = \tan\left(\frac{\phi_0}{2}\right)\left(\frac{E_0}{E}\right)\left[\frac{1}{\sqrt{\mathrm{RND}}} - 1\right] \tag{6.7}$$

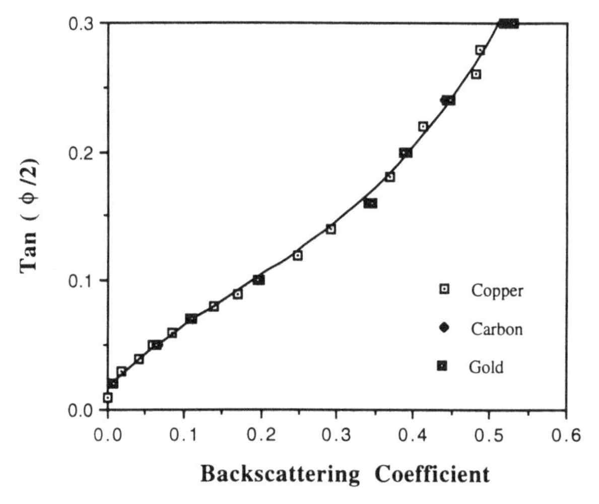

Figure 6.6. BS yield vs. tan ($\phi_0/2$).

Love et al. (1977) reasoned that since, from the form of the Bethe equation, the variation of (E_0/E) is substantially independent of atomic number, the backscattering coefficient η must depend only on the value of tan ($\phi_0/2$). As shown in Fig. 6.6, this assumption is correct. Using the code of Chap. 4, we find that the backscattering coefficient η depends on the value of tan ($\phi_0/2$) in a monotonic fashion but is to within statistical error independent of the atomic number and density of the target and of the electron energy. From a curve fit of the data in Fig. 6.6, we find the relation:

$$\tan\left(\frac{\phi_0}{2}\right) = 0.016697 + 0.55108\eta - 0.96777\eta^2 + 1.8846\eta^3 \qquad (6.8)$$

These coefficients differ slightly from those originally given by Love et al. (1977) and quoted in Eq. (4.7) because of the effect of the small angle-scattering correction included in our program [Eq. (4.10)]. To obtain the necessary estimate of η, we can now use the Hunger-Küchler relation [Eqs. (6.4) to (6.6)] discussed above. Given the atomic number of the target and the beam energy E, this gives us the value of η and hence of tan ($\phi_0/2$). This value can now be used in place of the original estimate [Eqs. (4.6) and (4.8)], with the advantage that the backscattering coefficient predicted by the Monte Carlo program will vary correctly with incident beam energy. It should be clear that this Hunger-Küchler-Love-Cox-Scott (HKLCS) model is not a radical revision of the plural scattering scheme but simply the replacement of Duncumb's one-parameter fit to the minimum scattering angle [Eqs. (4.4), (4.8), and (4.9)] with a more complex fit.

To incorporate this modification into the plural scattering code, it is only necessary to replace one procedure (hence the value of writing the code in a modular fashion). The revised procedure listed below replaces the original procedure of the same name and implements Eqs. (6.4), (6.5), (6.6), and (6.8). It is stored on the disk as HKLCS.PAS. To replace the old procedure with the new one using the TURBO PASCAL editor, go into EDIT mode and load PS_MC.PAS; then scroll through the program until the start of the procedure Rutherford_Factor. Type ⟨Control K R⟩ and a small window will appear asking for a file name. Type in HKLCS⟨return⟩. This block of code will then be read in from the disk and be placed at the cursor position. Now delete the old procedure and recompile the program.

```
Procedure Rutherford_Factor;

   {computes the Rutherford scattering factor for this incident ener-
   gy using the Love, Cox, Scott model and the Hunger-Küchler back-
   scatter equation}

var hkbs,hkc,hkl,hkm:extended;

  begin
        {compute the Hunger-Küchler backscattering coefficient}
              hkm:=0.1382-0.9211/sqrt(at_num);
               hkl:=ln(at_num);
               hkc:=0.1904-0.22236*hkl+0.1292*hkl*hkl
               -0.01491*hkl*hkl*hkl;
              hkbs:=hkc*power(inc_energy'hkm);

   {now compute the Love, Cox, Scott parameter from Eq. (6.8)}

rf:=0.016697+0.55108*hkbs-0.96777*hkbs*hkbs+1.8846*hkbs*hkbs*hkbs;
               rf:=rf*inc_energy;
  end;
```

Figure 6.7 plots the backscattering coefficient, computed using this version of the plural scattering program, for carbon, silver, and gold as a function of incident beam energy. The variation of η with energy is clearly evident below 10 keV, the yield for carbon rising as the energy falls while that for silver remains essentially constant and that for gold falls, ultimately to a value lower than that for silver. The predicted values agree well with the data plotted in Fig. 6.5 except at the lowest energies, where the Hunger-Küchler fit is probably inaccurate. Although this revised model is not a substitute for the more accurate Mott scattering cross section, it is often a sufficiently good approximation to make the use of the more complex Mott model unnecessary. An additional advantage of this approach is that, since an estimate for the backscattering coefficient is produced on the basis of the atomic number of the target from the Hunger-Küchler model, we could use Eq. (6.1)

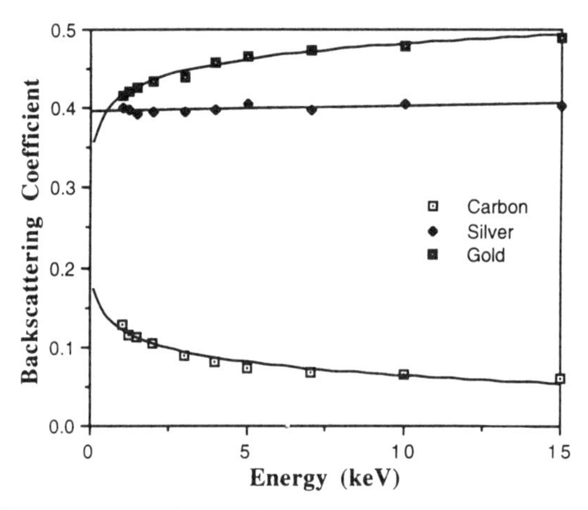

Figure 6.7. computed variation of BS yield with energy in HKLCS model.

directly to treat the case of a multicomponent material. Without coding this example in detail, the procedure would be as follows:

```
Get the number n_comp of components in the compound
    For i = 1 to n_comp
      Get atomic number Z[i]
       Get corresponding concentration c[i]
       Calculate BS coefficient ηi from Eqs. (6.4) through (6.6)
     Next i
   Calculate mean BS coefficient as ΣiCiηi
  Find Rutherford factor from equation 6.8
```

6.3 Predictions of the Monte Carlo models

We can now investigate some of the predictions that the Monte Carlo models make about the properties of backscattered electrons. These are parameters of the backscattering that are not, in either a direct or a hidden way, built into the models we have developed.

6.3.1 Variation of ξ with angle of incidence

The plural scattering models allow us to input the angle of incidence of the electron beam relative to the surface normal of the specimen. We can therefore calculate how the magnitude of the backscattering coefficient varies with the specimen tilt. Figure 6.8 shows some data for iron compared with an experimental measurement (Myklebust et al., 1976) for Fe—3.2% Si. In both cases the backscattering coefficient $\eta(\theta)$ at some angle θ is normalized by the corresponding backscattering coefficient at normal incidence $\eta(0)$, and the calculated and experimental data as obtained for

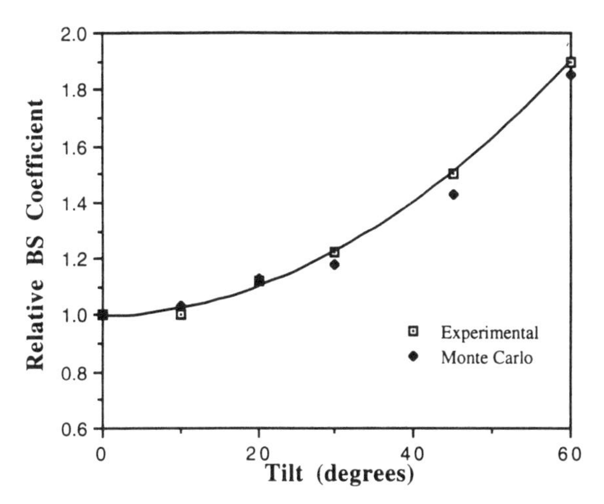

Figure 6.8. Experimental and computed variation of BS yield with tilt.

an incident energy of 30 keV. It can be seen that the amount of backscattering rises rapidly as the angle of incidence is increased, with both the experimental and the Monte Carlo data showing an increase of around 50% for a tilt angle of 45°, while for a tilt of 60°, the backscattering doubles in magnitude. These numbers will vary somewhat with both the incident beam energy and the atomic number of the target, although the general form of the $\eta(\theta)/\eta(0)$ variation remains fairly close to that illustrated in Fig. 6.8 except for the higher atomic numbers ($Z > 50$) and for low energies ($E < 3$ keV). The examination of such phenomena by running a series of simulations is an excellent way of becoming familiar with the use of a Monte Carlo model. Similar calculations can, of course, be performed with the single scattering model, but the relative slowness of this model makes obtaining a sufficiently good statistical accuracy a rather tedious affair. The actual data obtained are close to those of the plural scattering approximation.

6.3.2 Energy distribution and mean energy of backscattered electrons

The energy distribution of the backscattered electrons, or at least their average energy for a given set of conditions, is an important parameter, because the output from a backscattered detector depends on both the number of backscattered electrons (i.e., the value of η) and the energy of these electrons. Any measurement of the backscattered signal is therefore a convolution of the backscattering yield and the energy distribution from the sample, and consequently there will not be, in general, a simple relationship between this signal and the atomic number of the target. The energy distribution and mean energy can readily be found by modifying the `back_scatter` and `display_backscattering` procedures in the plural scattering model. We simply note on which step (1 to 50) of the trajectory the electron is traveling when it is backscattered and add 1 to the kth box of an array `bs_e[]`. Since the energy for all electrons on the k-th step is the same, i.e., $E[k]$, the average backscattering energy will be:

$$\text{mean\_energy} = \frac{\sum_{k=1}^{k=50} \text{bs\_}e[k]*E[k]}{\text{bk\_sct}} \tag{6.9}$$

because the total number of electrons backscattered is bk_sct. The modified procedures below implement these steps and plot the mean_energy value on a thermometer scale as a percentage of the incident beam energy.

Remember: add to the **var** list at the top of the program the entry:
bs_e:array[1 . . 50] of integer;

```
Procedure back_scatter;
   {handles special case of a backscattered electron}
  begin
        bk_sct:=bk_sct+1;       {add one to counter}
        num:=num+1;             {add one to total}
        bs_e[k]:=bs_e[k]+1;     {electron backscattered at k-th step}
        xyplot (y,z,yn,99);     {plot BS exit}
  end;

Procedure display_backscattering;
   {draws a thermometer to display the backscattering coefficient and
   another to indicate the mean energy of the backscattered elec-
   trons}

label     hang;
var       a,b,c,k:integer;
          mean_energy:extended;

  begin
               a:=GetMaxX-180;     {adjust to suit your screen}
               b:=bottom+23;       {ditto}
               c:=20;              {ditto}

               OutTextXY(a,b-8,"0 . . . . 0.25 . . . 0.5 . . . 0.75");
               OutTextXY(a+18,b+5,"BS coefficient");
               Line(a,b,trunc(a+(bk_sct/traj_num)*220),b); {draw it}

   {now compute the mean backscattering energy value}
                         mean_energy:=0; {initialize the value}
```

```
            for k:=1 to 50 do
               begin {add up number at step k*energy at step k}
                  mean_energy:=mean_energy+E[k]*bs_e[k];
               end;

                  mean_energy:=mean_energy/bk_sct; {averaged over
                  total BS}
                 {draw a thermometer to plot up this data value}

            OutTextXY(c,b-8,'0% . . . . . . . 50% . . . . . . 100%);
            OutTextXY(c+2,b+5,'% of incident energy');
            line(c,b,c+trunc(165*(mean_energy/inc
            _energy)),b);    {draw it in}

        hang:
                 if (not keypressed) then goto hang;    {freeze dis-
   play on screen}
            CloseGraph; {shut down the graphics unit}
end;
```

Figure 6.9 plots the value of mean_energy as a function of the atomic number for 15 keV incident electrons, showing that as Z increases, the value climbs steadily from carbon to gold. These values are seen to be in good agreement with the experimental data measured by Bishop (1966). From a curve fit of the Monte Carlo data, we can express the value of mean_energy as:

$$\text{mean\_energy} = E_0[0.55612 + 3.163*10^{-3}*Z - 2.0666*10^{-5}*Z^2] \quad (6.10)$$

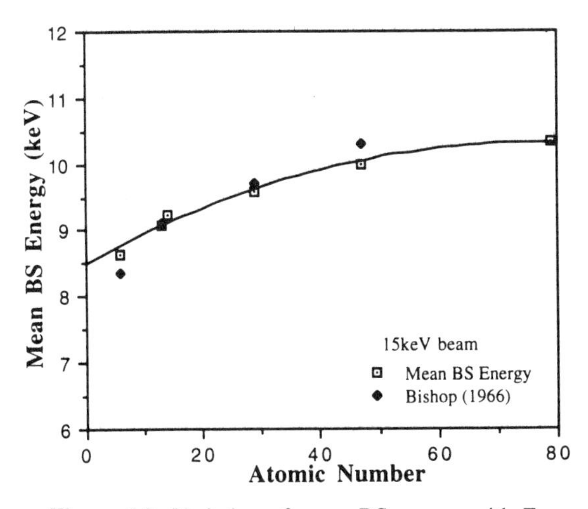

Figure 6.9. Variation of mean BS energy with Z.

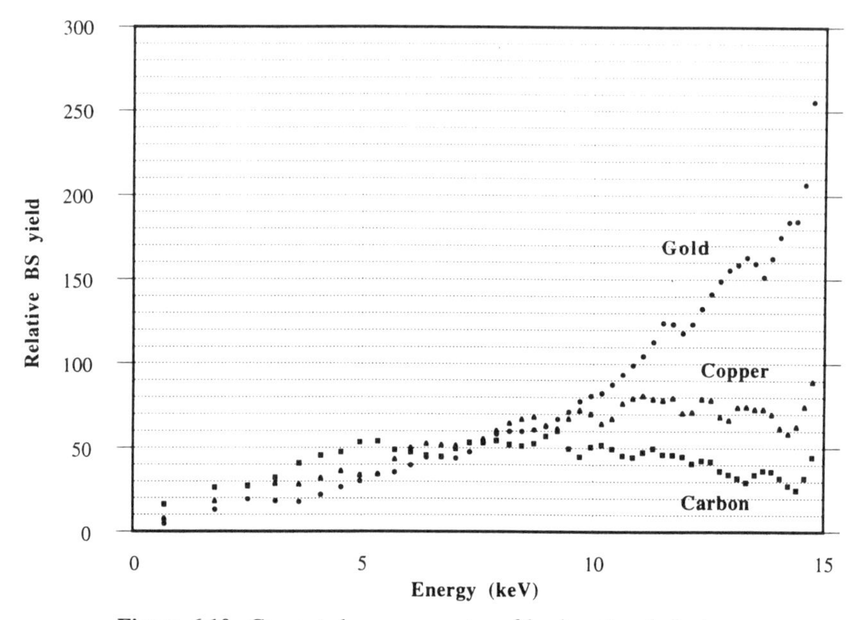

Figure 6.10. Computed energy spectra of backscattered electrons.

This variation is significant because backscattered detectors usually respond more efficiently to electrons of higher energy; therefore, as we move through the periodic table, the output from the detector will increase both because of the higher backscattering coefficient and because of the higher average mean energy. The reason for this variation in mean energy can be seen by plotting the histogram of the `bs_e[k]` array, computed above, which measures the number of backscattered electrons at each trajectory step k. As shown in Fig. 6.10, which plots data for $E_0 = 15$ keV beam energy, for carbon this histogram is more or less symmetrical about the energy $0.5E_0$; but for copper, the maximum in the curve has shifted upward in energy, while for gold the curve has no maximum but increases monotonically as the backscattered energy approaches the beam energy. The form of the energy distribution predicted by the Monte Carlo simulation is in generally good agreement with experimental data (Bishop, 1966; Mykelbust et al., 1976) except for the top end of the distribution close to the incident beam energy, where the number of backscattered electrons is overestimated somewhat. These particular electrons are those that are scattered in a single high-angle event close to the entrance surface, and the scattering of such electrons is accurately described only by the Mott cross section rather than by the screened Rutherford model used here. However, since these are only a small fraction of the total number of backscattered electrons, the error is insignificant.

Although we will not pursue it here, this same code fragment can also be used to estimate the information depth of the backscattered electrons—that is, the depth beneath the surface reached by an electron before it is backscattered. If an electron is backscattered on the k-th step, then the maximum possible depth to which it could have traveled beneath the surface is `k*step`, where `step` is the step length in the simulation, which is usually one-fiftieth of the Bethe range. An analysis of the `bs_e[k]` distribution therefore provides an upper bound on the information depth in the backscattered image. Typically this is about 0.3 of the Bethe range. The program JustBS on the disk incoporates this and the other modifications discussed above to give a detailed model of the backscattering effect, including an estimate of the surface area from which the electrons are emitted.

6.3.3 Variation of backscattering with density

In addition to comparing the predictions of our simulations with experimental data, we can use the Monte Carlo model to predict effects that might not be easily observable although they might also be important. An example of such an application is to ask the question: "How does the backscattering coefficient of a sample vary with density?" This question is of particular interest because of attempts to use the backscattered signal as a means of performing chemical composition analysis. While such a method might work for a homogeneous material, it seems natural to ask whether changes in the density of the target, caused perhaps by different processing methods, might not lead to a systematic error. Running the plural scattering simulation for copper at 15 keV, leaving all of the physical parameters unchanged except for the density and using 5000 trajectories for each value, produces the data shown in Fig. 6.11. To within the statistical error of the simulation, it can be seen

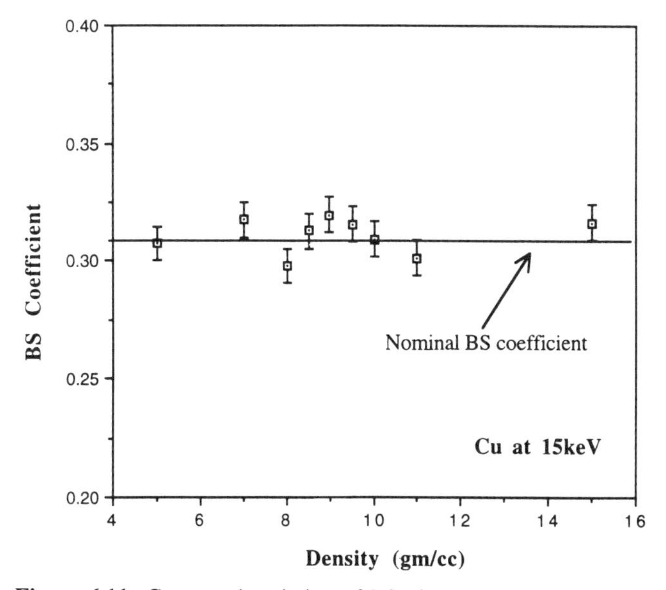

Figure 6.11. Computed variation of BS yield with sample density.

that changing the density over a wide range of values does not change the backscattering coefficient for an otherwise homogeneous material. We also find that the mean energy of the backscattered electrons, although not plotted, is independent of the density. While at first sight it might seem that this result is counterintuitive, an examination of the equations in Chaps. 3 and 4, which determine electron scattering, shows that the density appears only in the stopping power equation. Consequently varying the density changes only the Bethe range of the electrons, and this would not be expected to have any effect on η. If, however, the material is not homogeneous but contains regions of different density, changes might be expected; this situation is examined later in this chapter.

6.3.4 Variation of backscattering with thickness

A final demonstration of the models is to use them to compute how the backscattering coefficient might vary with the thickness of the target. Clearly, if the specimen is very thin, then the amount of backscattering is small; while for a sufficiently large thickness at a given incident energy, the backscattering coefficient should be constant. Our interest here is to simulate the variation of η between these two limiting conditions and to compare this with experimental data. It is immediately clear that, in this case, we cannot use the plural scattering approximation because, as discussed in Chap. 5, the granularity of this model is preset to be one-fiftieth of the Bethe range. When the target is a thin foil, each step of the plural scattering model would be a substantial fraction of the thickness of the foil and the quality of the approximation would be very poor. Instead, we must use the single scattering model, because the step length here is of the order of the elastic mean free path (e.g., a few nanometers at 10 keV), and is thus much smaller than the thickness being modeled. Figure 6.12 plots the predicted backscattering coefficient of films of carbon, copper, and gold as a function of their thickness at an incident beam energy of 15 keV. For convenience in comparison, the thicknesses in each case have been expressed as a fraction of the appropriate Bethe range R_B calculated from Eq. (4.1). The backscatter yield varies almost linearly with the specimen thickness (Niedrig, 1982) right up to the point where the backscattering yield reaches its "bulk" value. The thickness at which this occurs, however, varies significantly from one material to another, as would be expected from an examination of the interaction volumes pictures shown in Fig. 4.4. In carbon, the scattering of the electrons is weak, giving an interaction volume that hangs like a teardrop downward from the entrance surface. Consequently the specimen has to be almost 0.5 R_B thick before the backscattering saturates. In the case of gold, the electrons are strongly scattered, giving an interaction volume that is squashed up against the surface. As a result, the backscattering yield reaches its bulk value at only about 0.1 R_B. Since the Bethe range in gold is only a small fraction of that in carbon, this shows that the information depth of the

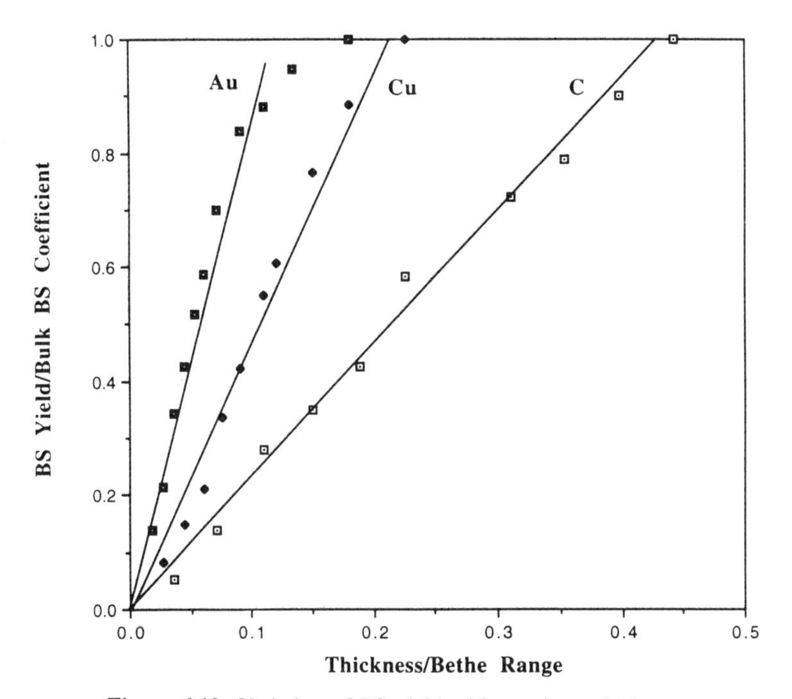

Figure 6.12. Variation of BS yield with specimen thickness.

BS signal, (i.e., the extent of the volume sampled by the backscattered electrons) is much less in gold than in carbon.

6.4 Modeling inhomogeneous materials

So far every application of Monte Carlo modeling we have considered assumes the specimen to be a infinite half-plane of uniform composition. In practice, of course, samples of such a restricted nature are of little interest, since in the real world of microscopy our specimens are finite in size; have edges, shape, and surface topography; and show wide variations of chemical composition from one point to another. Fortunately, Monte Carlo simulations, which track the electron step by step as it moves through the specimen, are ideally suited to this type of situation, so we can now start to develop the necessary tools to model such systems. In the simplest case, the sample might consist of a layer of material 1 on top of a substrate of materials 2. If the depth of the interface between these materials occurs at a depth `boundary`, we can then determine which material the electron is in by testing the value of `z`.

```
{the sample consists of two materials i=1 and i=2}
          i:=1;    {reaches to player first}
      if z>=boundary then i:=2;
```

The program below implements this example of a thin film on a substrate to illustrate how the original program of Chap. 4 is modified to deal with two materials. In order to save space, functions and procedures that have been discussed already, either in this chapter or earlier chapters, are not listed but simply identified as required. A further generalization to three or more materials, or to the situation where a single scattering model is required, is a simple matter; a program of this type is given in the chapter on x-ray generation.

```
Program BINARY;
   {this code performs a plural scattering Monte Carlo trajectory
   simulation for the case where a layer of material A is placed on
   top of a bulk substrate of some other material B. There is no
   graphical display in this program}

   {$N+}   {turn on numeric coprocessor}
   {$E+}   {install emulator package}

uses    CRT,DOS;              {resources required}

label       back_scatter,abort;

const       two_pi=6.28318;    {2 constant}

var
   inc_energy,boundary,tilt,s_tilt,c_tilt:extended;
   nu,sp,cp,ga,an,an_m,an_n:extended;
   x,y,z,xn,yn,zn,ca,cb,cc,cx,cy,cz,v1,v2,v3,v4:extended
   E:array[1 . . 2,1 . . 51] of extended;
   at_num:array[1 . . 2] of extended;
   at_wht:array[1 . . 2]of extended;
   mn_ion_pot:array[1 . . 2] of extended;
   m_t_step:array[1 . . 2] of extended;
   density:array[1 . . 2] of extended;
   step:array[1 . . 2] of extended;
   rf:array[1 . . 2] of extended;
   bk_sct,num,traj_num,i,k:integer;

Function power(mantissa,exponent:real):extended;

   Function stop_pwr(i:integer;energy:extended):extended;
   {calculate stopping power for material i using Eq. (3.21)}
var temp:extended;
```

```
  begin
     if energy<0.05 then {to avoid problems as energy goes to zero}
           energy:=0.05;
           temp:=ln((1.166*energy+0.85*mn_ion_pot[i])/mn_ion_pot[i]);
        stop_pwr:=7.85E4*at_num[i]*temp/(at_wht[i]*energy);
  end;

Procedure range(i:integer);
   {calculates range assuming Bethe continuous energy loss}

var energy,f,fs,bethe_range:extended;
         l,m:integer;

  begin
          fs:=0.;           {initialize variable to be sure}
      for m:=1 to 21 do     {Simpsons rule integration}
          begin
             energy:=(m-1)*inc_energy/20;

               f:=1/stop_pwr(i,energy);

             l:=2;
                if m mod 2=0 then l:=4;
                if m=1 then l:=1;
                if m=21 then l:=1;
                    fs:=fs+l*f;
          end;

   {now use this to find the range and step length for these condi-
   tions}

                bethe_range:=fs*inc_energy/60.0; {in g/cm2}
                m_t_step[i]:=bethe_range/50.0;
                bethe_range:=bethe_range*10000.0/density[i]; {in mi-
                crons}

      GoToXY(40*(i-1)+1,13); {display 1 on LHS, 2 on RHS of screen}
        writeln('Range in' i, 'is bethe_range:4:2' 'microns');

                step[i]:=bethe_range/50.0; {unit step of simulation}
  end;
Procedure profile(i:integer);
   {compute 50-step energy profile for electron beam}

var A1,A2,A3,A4,em:real;
                  m:integer;
  begin
```

```
                E[i,1]:=inc_energy;
            for m:=2 to 51 do
      begin

            A1:=m_t_step[i]*stop_pwr(i,E[i,m-1]);

            A2:=m_t_step[i]*stop_pwr(i,E[i,m-1]-A1/2);

            A3:=m_t_step[i]*stop_pwr(i,E[i,m-1]-A2/2);

            A4:=m_t_step[i]*stop_pwr(i,E[i,m-1]-A3);

           E[i,m]:=E[i,m-1] - (A1 +2*A2 +2*A3 +A4)/6.;

      end;
                          E[i,51]:=0;

   {now smooth these profiles out a little bit}

            for m:=2 to 50 do
              begin
                E[i,m]:=(E[i,m] + E[i,m+1])/2.;
              end;

 end;

Procedure Rutherford_Factor(i:integer);
   {find the screened Rutherford scattering parameter b using the
   HKLCS method}
var hkbs,hkc,dum,hkm:extended;
 begin
         hkm:=(0.1382 - 0.9211/sqrt(at_num[i]));
            dum:=ln(at_num[i]);
         hkc:=0.1904-0.2236*dum+0.1292*dum*dum-0.01491*dum*dum*dum;
         hkbs:=hkc*power(Inc_Energy,hkm);

        {then the scattering factor for material i is}

   rf[i]:=0.016697+0.55108*hkbs-0.96777*hkbs*hkbs
+1.8846*hkbs*hkbs*hkbs;
 end;

Procedure set_up_screen;
   {gets necessary input data to run the program}
 begin
         ClrScr;   {tidy up the display screen}
```

```
      GoToXY(25,1);
       Writeln('*A on B MC Simulation in Turbo Pascal*');
{Having set up the screen now get input data}

   GoToXY(1,5);
    Write('Input beam energy in keV');
      Readln(Inc_Energy);

   GoToXY(1,7);
    Write('Boundary depth (microns)');
      Readln(boundary);

   GoToXY(1,9);
    Write('Tilt angle in degrees');
      readln(tilt);
           s_tilt:=sin(tilt/57.4);    {convert degrees to radians}
           c_tilt:=cos(tilt/57.4);
    ClrScr; {again-to set up for two column input of data}

  for i:=1 to 2 do {get materials data}
   begin
          if i=1 then    {surface layer}
             begin
              GoToXY(1,5);
               writeln('Surface Layer');
             end
          else              {we are in the substrate}
             begin
              GoToXY(40,5);
                 writeln('Substrate');
             end;
 {now get the materials information that is needed}
   GoToXY(40*(i-1)+1,7);    {LHS of screen for i=1, RHS for i=2}
    Write('. . . . . Atomic Number is');
      Readln(at_num[i]);
    GoToXY(40*(i-1)+1,9);
      Write('. . . . . Atomic Weight is');
       Readln(At_wht[i]);
    GoToXY(40*(i-1)+1,11);
      Write('. . . . . . density in g/cc is');
       Readln(Density[i]);

 {Calculate the Berger-Selzer Mean ionization potential mn_ion_pot}

    mn_ion_pot[i]:=(9.76*at_num[i] + (58.5/power(at
_num[i],0.19)))*0.001;
```

```
        {now get the Rutherford factor b using the HKLCS method}
                        Rutherford_factor(i);

   {now get the range and step length in microns for these conditons}

                              range(i);
   {and calculate a 50-step energy profile E[i,k] using this result}

                              profile(i);

    end;    {of the input loop}

            {get the number of trajectories to be run in this simula-
            tion}

    GoToXY(18,15);
     write("Number of trajectories required");
       readln(traj_num);

                {and set up the display for output of data}

                          GoToXY(1,16);
writeln('---------------------------------------------------------');
                          GoToXY(1,17);
                    writeln('Number of trajectories');
                          GoToXY(40,17);
                    writeln('Backscattered fraction');
  end;

Procedure init_counters;

Procedure reset-coordinates;

Procedure p_scatter(i:integer);
   {calculates scattering angles using plural scattering model of
   Chap. 4}

  begin
    {call the random number generator function}
                          nu:=sqrt(RANDOM);
                          nu:=((1/nu)-1.0);
                          an:=nu*rf[i]*inc_energy/E[i,k];

    {and use this to find the scattering angles}
                          sp:=(an+an)/(1+(an*an));
                          cp:=(1-(an*an))/(1+(an*an));
```

```
    {and the azimuthal scattering angle}
                              ga:=two_pi*RANDOM;

  end;

Procedure new_coord(i:integer);
   {calculates new coordiantes xn,yn,zn from x,y,z and scattering an-
   gles}
  begin
             {the coordinate rotation angles are}
                              if cz=0 then cz:=0.000001;
                              an_m:=-cx/cz;
                              an_n:=1.0/sqrt(1+(an_m*an_m));

   {save computation time by getting all the transcendentals first}
                              v1:=cos(an_m)*sp;
                              v2:=cos(an_n)*sp;
                              v3:=cos(ga);
                              v4:=sin(ga);
   {find the new direction cosines}
                    ca:=(cx*cp) + (v1*v3) + (cy*v2*v4);
                    cb:=(cy*cp) + (v4*(cz*v1 - cx*v2));
                    cc:=(cz*cp) + (v2*v3) - (cy*v1*v4);
   {and get the new coordinates-using the appropriate step}
                    xn:=x + step[i]*ca;
                    yn:=y + step[i]*cb;
                    zn:=z + step[i]*cc;

  end;

Procedure reset_next_step;

**************************************************
*        this is the start of the main program         *
**************************************************

                                set_up_screen;
                                init_counters;
                                 randomize; {reseed random number
                                 generator}
{

**************************************************
*                  the Monte Carlo loop                *
**************************************************
```

```
}

                        while num < traj_num do
  begin
                           reset_coordinates;

                         for k:=1 to 50 do
       begin
     {first find out where the electron is now i=1 or 2. In the
      simplest case this is done by checking whether or not the
       electron z coordinate places it below the boundary layer}
                 if z>=boundary then i:=2
                     else
                         i:=1;
   {in other situations a different test would be inserted here}

                  {now allow the electron to be scattered}
                          p_scatter(i);
                          new_coord(i);
     {test for electron position within the sample}
                 if zn<=0 then
                 begin
                   bk_sct:=bk_sct+1;
                   num:=num+1;
                   goto back_scatter;
                 end
                 else
                    reset_next_step;

  end;   {of the 50-step loop}

               num:=num+1;   {add one to the trajectory total}
             back_scatter;   {end of goto branch for BS}
          {update the screen displays}
                   GoToXY(25,17);
                   writeln(num);   {display trajectory total}
                   GoToXY(65,17);
                   writeln((bk_sct/num):4:3);   {display BS
                   coefficient}
          {to escape from the program press any key}
                   if keypressed then goto abort;
{

**************************************************
*             end of the Monte Carlo loop        *
**************************************************
```

```
}
        end;    {of the Monte Carlo loop}
                abort:    {escape from the program}
                 GoToXY(21,21);
                    writeln('. . . htat's all folks');
                        readln; {freeze the display}

 end.
```

6.5 Notes on the program

The program starts in the same way as for the previous examples by setting up the pragmas that select the use of the math coprocessor or the emulator package. Since this program does not generate any graphical display, the `uses` declaration includes only the calls for the functions CRT and DOS, which are required. The variable declaration section resembles that used before but with an important difference. Instead of there being a single value of, for example, the atomic number `at_wt`, there will now be one of two possible values depending on where within the target the electron is situated. Therefore the original single value of `at_wt` is replaced by an array of `at_wt` values. In the notation used in PASCAL, we indicate this by writing `at_wt [i]` to represent the *i*-th value of `at_wt`. We must thus modify the list of variables in the program to indicate that `at_wt`, `density`, or any of the other specimen parameters can take two (or more general examples *n*) different values. The `VAR` table would then contain entries such as:

```
at_wt:array [1 . . 4] of extended;
at_num:array [1 . . 4] of extended;  {and so on for every variable
that changes}
```

where the [. . .] brackets indicate the number of the first and last members of the array and specify what type of variable it is. Note that in PASCAL we must specify an actual number of members of the array; this cannot be done in terms of another variable. Having now done this, we must also indicate to the various `PROCEDURES` and `FUNCTIONS` that use these variables which of the *n* values they are to use. We do this by passing a parameter to the procedure, which tells it which particular value to use. In each case, where in Chap. 4 we had a single-valued variable (e.g., `density`), we now have one specific value taken from the array of possible values `density`[*i*[. The value of *i* to be used is passed to the procedure or function at the time that we call it; therefore, instead of calling `profile` to calculate the energy profile $E[k]$, we call `profile`[*i*] to tell it to calculate the profile for the *i*-th material by substituting the appropriate values of `stopping_pwr`, `m_t_step` etc. One assumption in this program is that $E[1,k] \approx E[2,k]$; in other words, that the instantaneous energy of the electron depends only on the number of the trajectory step k and not on the actual material. This approximation is, in fact, closely correct, because—from the form of the Bethe equation—the instantaneous energy depends only on the fraction of the Bethe range traveled and not significantly on the atomic number of the material itself.

The rest of the program then follows closely on the original version given in Chap. 4 except that we now test at the top of the loop which of the two materials the electron is in, $i=1$ or $i=2$, before computing the scattering angles and finding the new coordinates. Because we are not trying to display these data visually, the procedures for graphics setup and plotting are missing, and if you run this program, you will notice how much faster it performs than the earlier version. While pictures of the trajectories are interesting and quite often helpful, they are very time-consuming to produce because of the large time overhead required in calculating plotting coordinates and drawing the screen displays. Therefore they should be omitted if not strictly necessary.

Figure 6.13 shows an application of this program to the case of a thin carbon film on top of a gold substrate at 10 keV. Since we are using a plural scattering model here, the data will not be reliable for film thicknesses less than about a thousand angstroms (i.e., three or four trajectory steps), but the trend is clear (Hohn et al., 1976). At low film thicknesses, the backscattering coefficient is essentially that of gold; this then decreases with film thickness, eventually falling to the value

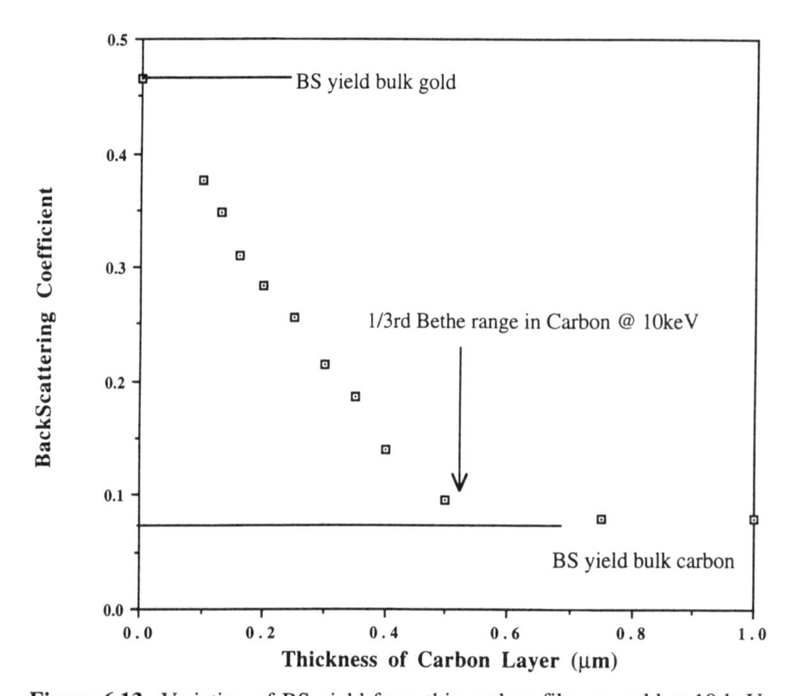

Figure 6.13. Variation of BS yield from thin carbon film on gold at 10 keV.

appropriate for bulk carbon. The film thickness at which this occurs is about the same as that for which an unsupported thin film of carbon attains its bulk scattering value (see Fig. 6.12). This thickness is an estimate for the "depth of information" in the backscatter image at this beam energy (i.e., the maximum distance beneath the surface at which we could deduce—from information in the BS image—that the sample was not solid carbon). Several other applications of this type of multilayer program are considered later in this book—for example, in the chapters on electron beam induced charge (EBIC) and x-rays.

Depending on the problem, it is sometimes adequate to test the end point `zn` of a trajectory step to see which part of the specimen it is in, but in other cases cases it may be better to find the midpoint of the step and test that instead:

```
z_mid:=(z+zn)/2;
if z_mid>=depth then i:=2;
```

The first approach is quicker, but the second makes more use of the data in the simulation and is less likely to cause problems and confusion when complicated geometries are being considered. Care must also be taken to properly error-trap the test function; for example, if the electron is backscattered `zn<0`, then the resultant value of `z_mid` may be negative and lead to a problem. The safest procedure is to deal with the special cases of backscattering and transmission before other conditions are tested for. Finally, in any case where the regions being considered in the problem are small in size compared to the Bethe range, the single scattering model must be used to avoid errors.

For complicated geometries, the problem of specifying the region of the sample i in which the electron is moving also becomes more complex because it will involve all three of the coordiantes rather than just the z axis. The obvious way to tackle the problem is to apply a string of suitably structured `if . . . then` tests to make the decision. For example,

```
if x_mid<right_hand_edge and y_mid>back_side
              then i:=2 else
          if z>bottom then i:=3
                else i:=1;
```

etc. but this is cumbersome and slow, as these tests have to be made on every step of every trajectory. For a completely arbitrary geometry, however, such as a random steps on a surface, this is the only available way. One alternative is to divide up the entire volume of interest into an array of cubes of a size comparable with or smaller than the expected step length and then to apply the sort of test described above to preassign a value of i to each element of the array and store it, so that, given the coordinates of the electron, the correct value of i can simply be looked up. This is fast and quite general but requires very large arrays of data in all except the simplest cases (Joy, 1989a).

In one special case, however, there is a simple and elegant solution to this difficulty. If each of the regions of interest in a material is a closed volume that can be described by an equation of the form $f(x, y, z) = 0$, then we can immediately determine whether any given point p, q, r is inside, on the surface of, or outside the volume by evaluating $f(p, q, r)$. If $f(p, q, r) < 0$, then the point is inside the volume; if $f(p, q, r) = 0$, then the point is on the surface; and if $f(p, q, r) > 0$, then the point is outside. For example consider the case of a sphere, radius r and centered at the coordinates $(0, 0, Zc)$. We can find whether or not the point (x, y, z) is inside or outside the sphere by a simple PASCAL function, which can be called simply `Outside`.

```
Function outside(xv,yv,zv:extended):Boolean;
   {tests whether or not the electron at xv,yv,zv is inside or out-
   side the sphere of radius r centered at 0,0,Zc}
  var dum:extended;
  begin
       dum:=xv*xv + yv*yv + (zv-Zc)*(zv-zc) - r*r;
            if dum<=0 then outside:=false
                  else
               outside:=true;
  end;
```

The start of the Monte Carlo loop in the program above would then be modified to read

```
                              while num < traj_num do
  begin
                                 reset_coordinates;

                               for k:=1 to 50 do
        begin
   {first find out where the electron is now—i=2 inside sphere
   or i=1 outside the sphere}

     if outside (x,y,z) then i:=1
         else
             i:=2;
  etc.
```

remembering that an `if.⟨test⟩ . . then` statement is carried out if the test is true (or is a function that evaluates to a positive number).

Figure 6.14 shows some data obtained from this version of the program. The

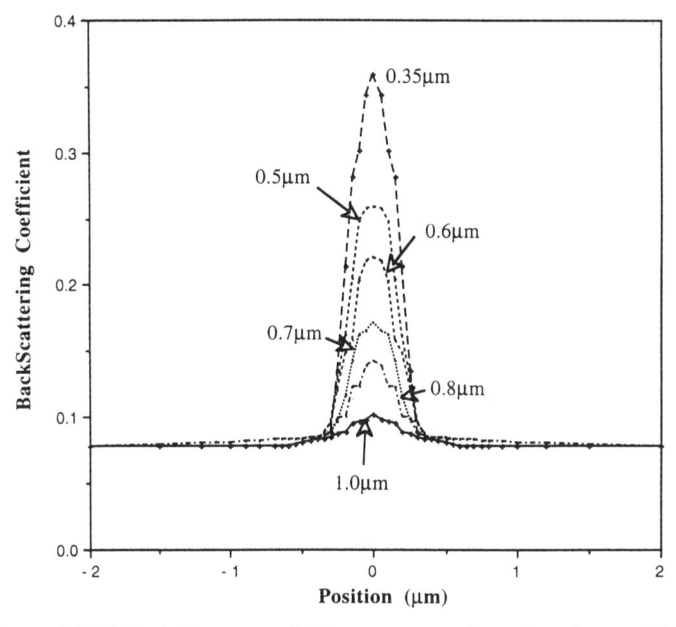

Figure 6.14. BS yield across a 0.25-µm copper sphere in carbon at 15 keV.

matrix of material was carbon, containing a copper sphere 0.25 µm in radius, and the incident beam energy was 15 keV. The program was set to calculate the variation of η with the position of the beam relative to the center of the sphere—which, because of the cylindrical symmetry of the problem, we need only do along a single radius. An additional loop was therefore inserted into the program to move the value of the starting y coordinate of the beam from 0 to some maximum value `y_max.` in suitable steps. The data of Fig. 6.14 show how the apparent backscattering coefficient varies with the position of the beam and the depth of the sphere. We see that for $Zc > 1.0$ µm, or about one-third the Bethe range, the sphere is not visible, so the "depth of information" in this case is about 1 µm. As the sphere is brought closer to the surface, its visibility increases and its apparent width changes, the full width at half maximum of the profile being 0.25 µm when the sphere is at a depth of 0.7 µm but 0.4 µm wide when the sphere is at 0.35 µm depth. These values for the apparent width of the image are interesting because they indicate the sort of spatial resolution that can be achieved in the backscattered image mode. On a simple basis, we might expect that the apparent size would be of the order of the actual diameter of the sphere plus a term to account for the beam interaction volume. However, as we see, the measured size is always less than the actual size. This is because the mere fact that an electron reaches the copper sphere does not guarantee that it will be backscattered by it, since in order to escape, the electron must first travel through a significant amount of the carbon matrix. When the sphere is deep or on the periphery of the interaction volume, the electrons reaching it are already low in energy and have little chance of reaching the surface even if scattered directly toward it.

This method of testing the position of the beam relative to a closed surface is certainly a special case, but it is not as restrictive as it might at first appear, because a sphere can be generalized into an ellipse (i.e., the equation becomes of the form $ax^2 + by^2 + cz^2 - d$), which can then, by an appropriate choice of major and minor axes, be turned into a disk or a plate in any chosen orientation. A search through books on analytical geometry will also reveal a few other closed surfaces (or effectively closed, as in the case of a hyperbola) that might also prove useful.

A case of some practical importance occurs when the surface encloses not another material but a vacuum—i.e., when the material contains a void. As can readily be confirmed, this situation cannot be taken care of by expedients—such as representing the void as being a material of zero atomic number and density—because the stopping power and scattering expressions do not behave properly in this limit. Instead, we assume that an electron entering the void will neither be scattered nor deposit significant energy but will simply move in a straight line across the void until it once again leaves. Using the notation from the code given above, the program would be modified to read

```
      while num < traj_num do
begin
            reset_coordinates;
            for k:=1 to 50 do
      begin
  {first find out where the electron is now—i=2 is inside the void
        or i=1 outside in the matrix"

          if outside(x,y,z) then i:=1
            else
              begin {extrapolate previous trajectory step}
                go_round_agin:{label for next loop if needed}
                  xn:=x+step[1]*cx;
                    yn:=y+step[1]*cy;
                  zn:=z+step[1]*cz;
                if outside (xn,yn,zn) then   {its gone
                                  through}
                                  go to escaped   {so leave
                                  this loop}
                    else {extend the extrapolation}
                      x:=xn;y:=yn;z:=zn;    {reset
                      coordinates}
                go to go_round_again;          {and do it
                again}
              end;
```

```
        escaped: {label for exit from loop}
            i:=1; {remember to reset this on exit}

etc.
```

While the programming style will win no prizes because of its use of go-to statements, this fragment of code correctly handles the void. The visibility and apparent size of voids behaves in an approximately similar way to that for the inclusions discussed above, except that the absence of a scattering material leads to a fall in the backscattering coefficient rather than to a rise. For an interesting study of contrast from voids in the scanning electron microscope (SEM) using Monte Carlo techniques of this type, see Gasper and Greer (1974).

6.6 Incorporating detector geometry and efficiency

A final topic that must be discussed concerns the difference between the computed backscattering coefficient and the actual signal collected in the SEM by a suitable backscatter detector. If the backscatter detector were perfect, that is, if it collected every backscattered electron within the 2π steradians above the sample surface and if every electron—regardless of energy—produced the same magnitude of output pulse, the signal would be directly proportional to the backscattering coefficient. In general, this is not the case, and circumstances usually require that the properties of the detector be taken into account when a Monte Carlo simulation is being made. As a rule, the position of the detector must be considered if the specimen to be modeled has anything other than a flat (i.e., planar) surface and, since all practical backscatter detectors have an output that varies directly with the energy of the electrons collected, modeling the efficiency of the detector for electrons of different energy is always likely to be useful.

In order to incorporate the geometrical properties of the detector into the program, it is only necessary to check whether or not any given backscattered electron is traveling in the right direction to impinge on the detector. If the detector is assumed to be annular about the incident beam—if the outer diameter of the detector is R_o and the distance from the detector to the sample is D—then, with the coordinate convention being used in this book, an electron will be collected if its direction cosine `cc` satisfies the relation

$$\mathrm{cc} <= -\cos\{\arctan(R_o/2D)\} \tag{6.11}$$

note that the sign is negative because the positive z-direction is defined to be in the beam direction. The code of Chaps. 4 and 6 would then be modified to read:

```
if zn<=0 then
  if cc<=-critical_angle then
```

```
begin
  bk_sct:=bk_sct+1;
  num:=num+1;
  goto back_scatter;
end
```

where critical_angle is set equal to the quantity on the right hand side of Eq. (6.11). In this way the characteristics of a specific backscattered detector can be added to the simulation by feeding in the values for R_o and D at a suitable point in the program. If the detector is annular but split into two halves or four quadrants, the values of the other direction cosine components must also be considered. For example,

```
if zn<=0 then
          begin
            if ((cc<=-critical angle) and (cb>=0)) then
               detector_A_signal:=detector_A_signal+1
                   else
               detector_B_signal:=detector_B_signal+1;
            num:=num+1;
            goto back_scatter;
          end
```

where for this split detector we keep separate track of the electrons collected on the two halves A and B (remembering first to declare and initialize these variables). The extension of this code to other more complex arrangements is obvious.

All backscatter detectors display some energy sensitivity. Typically, we find that their is some minimum energy E_{min} below which the backscattered electron creates no output signal, while above E_{min} a backscatter electron of energy E will produce an output signal that varies linearly as $(E-E_{min})$. The code for our plural scatter models would now look something like this:

```
if ((zn<=0) and (E[i,k]>E_min)) then
          begin
            if ((cc<=-critical angle) and (cb>=0) then
             detector_A_signal:=detector_A_signal+(E[i,k]-E_min)
                   else
             detector_B_signal:=detector_B_signal+(E[i,k]-E-min);
            num:=num+1;
            goto back_scatter;
          end
```

so that the detected signal now depends on the actual energy of the electron collected. Note that the variables detector_A_signal and detector_B_signal must

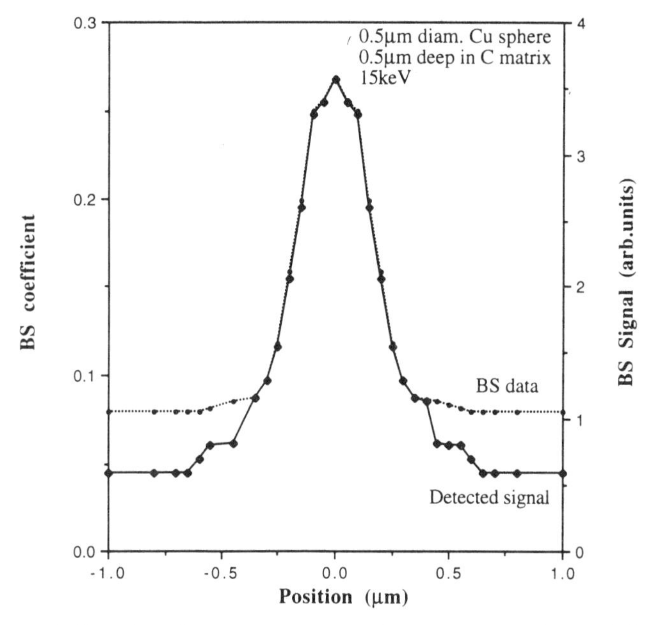

Figure 6.15. BS profile across sphere and corresponding detected signal.

now be declared as `extended` or `real` rather than integer and that the final result will no longer represent the backscatter yield but instead a scaled representation of the actual signal detected. Figure 6.15 shows the effect that this kind of correction has. The datum is one of the traces from Fig. 6.14, but now corrected for a detector that has an E_{min} value of 5 keV. The two traces show the variation of the backscattered coefficient and the output signal from the backscatter detector respectively. While the forms of the profiles are essentially identical, note that the effect of including the energy sensitivity of the detector is to increase the apparent visibility of the copper sphere, since the "peak-to-background" variation of the signal is increased by about 20%. There is thus no longer a direct proportionality between the backscatter coefficient and the detected backscatter signal. This is of importance if the backscatter signal is being used for chemical observations, since it cannot be assumed that the detected signal and the emitted backscattering coefficient—and hence the mean atomic number of the target—are all simply related together. Because the detector responds as much to changes in the energy of the electrons it receives as it does to their actual number, these two effects convolute together and cannot simply be separated. Quantitative backscattered imaging is therefore a difficult activity unless care is taken to account for all of these factors (e.g., Sercel et al., 1989).

7

CHARGE COLLECTION MICROSCOPY AND CATHODOLUMINESCENCE

7.1 Introduction

Charge collection imaging in the scanning electron microscope, often known by the acronym EBIC (electron beam–induced current), has become a widely used technique for the characterization of semiconductor materials and devices (Leamy, 1982; Holt and Joy, 1990). While there is a substantial literature on the use of EBIC methods to measure semiconductor parameters, such as the minority carrier diffusion length (Leamy, 1982), the majority of charge-collected images are interpreted in a purely qualitative manner. Cathodoluminescence (C/L) imaging of semiconductor materials, which is in essence very similar to EBIC, has similarly been used mostly in a picture-taking rather than a data-producing mode. The problem is not that there are no good models to explain the image formation but rather that, in order to provide tractable analytical expressions for the calculation of contrast, it is invariably necessary to make significantly oversimplified assumptions about the interaction of the electron beam with the specimen. In this chapter, we demonstrate how the Monte Carlo models discussed earlier can be used to overcome these problems and make EBIC and C/L more useful techniques for microcharacterization.

7.2 The principles of EBIC and C/L image formation

When an electron beam impinges on a semiconductor, some of the energy deposited by the beam is used to promote an electron from the filled valence band, across the band gap, to the empty or partially filled conduction band (Fig. 7.1). Since the valence band was initially full, the removal of an electron leaves behind a "hole" that has all the physical properties of an electron but carries a positive charge. For each electron promoted across the band gap, one hole is formed, so it is convenient to consider the two components together and talk of an electron-hole pair. To generate the electron-hole pair requires an amount of energy e_{eh}. Typically, e_{eh} is about three times the energy of the band gap (Table 7.1); so, for example, for silicon, e_{eh} is 3.6 eV. If we assume that all of the energy deposited by the incident beam of energy E_0 is ultimately unavailable for the generation of electron-hole pairs, then the number of carrier pairs formed n_{eh} will be

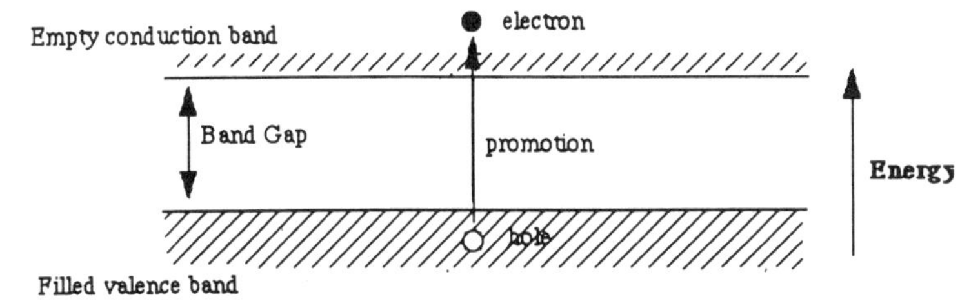

Figure 7.1. Electron-hole pair generation in a semiconductor.

$$n_{eh} = E_0/e_{hh} \tag{7.1}$$

That is, a 10-keV beam incident on silicon could produce 10,000/3.6 ≈ 2800 electron-hole pairs. Because the electron and hole have opposite charges, they are electrostatically attracted and will tend to drift together through the lattice, maintaining local electrical neutrality. After only a few femptoseconds (10^{-15} s), the electron will fall back across the band gap and recombine with the hole, giving up, as it does so, some of the energy used to form the original carrier pair. It is this effect that may result in the production of C/L emission, since the excess energy can be carried away by an emitted photon. In any event, within a very short time after the passage of the incident electron, the system has returned to its original state.

Since, unlike a metal, a semiconductor has significant resistivity, a potential difference can be maintained across it, resulting in the generation of an electric field imposed across the sample (Fig. 7.2A). An incident electron of energy E_0 will, as before, produce some number of carrier pairs n_{eh}, but now, because the electrons and holes carry opposite charges, the electrons will tend to move toward the positive end of the sample while the holes will drift toward the negative end. One result of this is

Table 7.1 Electron-hole pair energies

Material	Electron-hole pair energy (eV)
C (diamond)	13.1
CdS	7.5
CdTe	4.8
GaAs	4.6
GaP	7.8
Ge	3.95
InP	2.2
PbS	2.0
Si	3.6
SiC	9.0

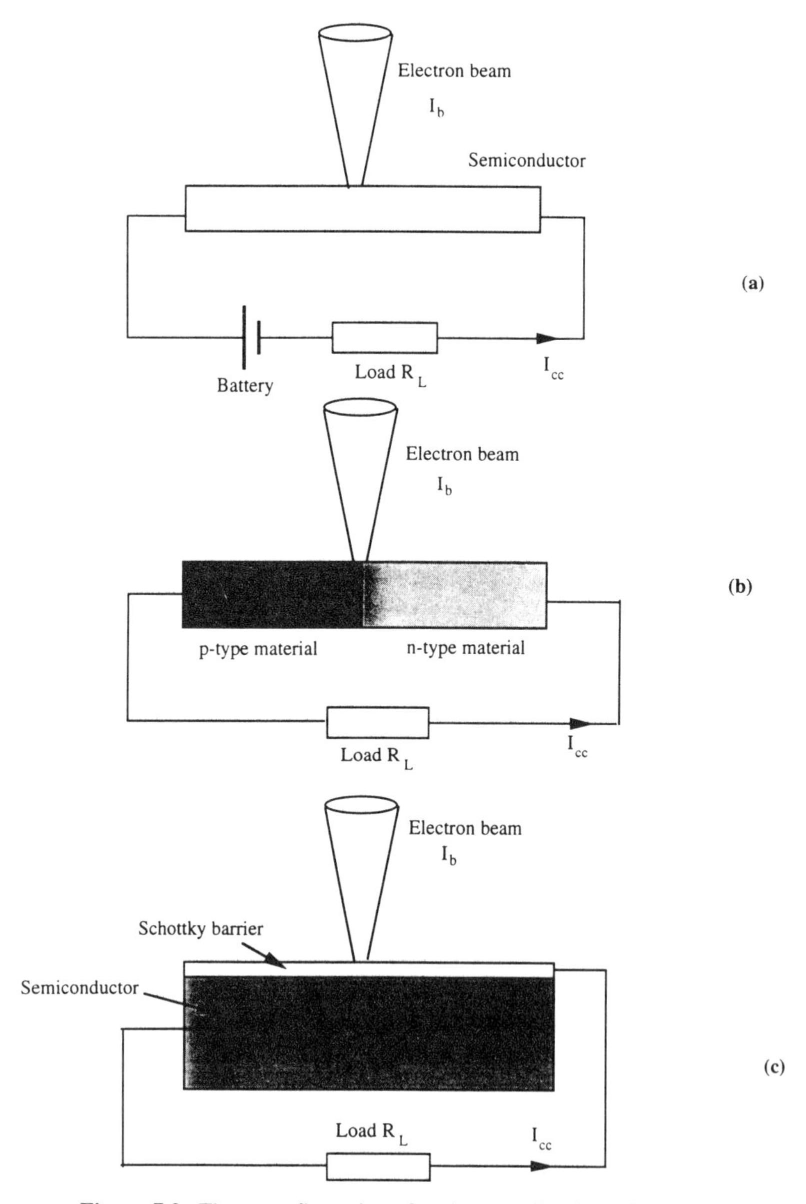

Figure 7.2. Three configurations for charge collection microscopy.

immediately apparent. Before the electron beam is turned on, the amount of current flowing through the semiconductor is very small or zero, because there are no electrons in the conduction band and hence there is no way to move charge. But with the beam striking the specimen, each electron produces n_{eh} electron-hole pairs, and the presence of these free carriers will permit charge to flow. Thus the electron beam has induced conductivity in the semiconductor. This effect is known as electron beam–induced conductivity (and so gives another interpretation of the acronym EBIC) or β-conductivity (Holt and Joy, 1990). If the beam is scanned, the resultant conduction signal can be used to form an image.

In practice, a more interesting procedure is to generate the field internally within the semiconductor. For example, if a semiconductor that has been chemically doped to make it a p-type material (i.e., one containing an excess of holes) is placed in contact with n-type material (i.e., one with an excess of electrons), then a region is formed around the p-n junction where a potential and hence a field is present (Fig. 7.2B). This field arises from the attempt of the electrons in the n-type material to go to the p-type region and the holes in the p-type material to flow into the n-type region. Within a short time, charge from uncovered atomic nuclei produces a field just sufficient to prevent any further charge motion. In this region of field, there can be no free charge carriers, so it is called a "depleted region," and it typically extends for a few micrometers on either side of the physical location of the p-n junction (Leamy, 1982). Alternatively, if a thin metal film is put into atomic contact with the surface of a semiconductor (Fig. 7.2C) then this "Schottky barrier" again results in the appearance of a depleted region extended downward from the Schottky layer. Without an incident electron beam, there exists, in either case, a static potential between the p and n regions or between the metal and semiconductor; but if the p- and n-type regions, or the Schottky barrier and the semiconductor, were connected together by a wire, then no current would flow, because the field exists only in a region from which all of the available charge carriers have already been removed.

Now let us place the incident electron beam onto the p-n specimen in either the p- or n-type region but away from the depleted zone. As before, electron-hole pairs will be generated, but since the material in which they are produced is electrically neutral and has no field across it, they will quickly recombine and no external effect will be observed. If, however, the beam is placed in the depleted region, the carriers produced will see the depletion field and the carrier pairs will be separated. This motion of charges within the specimen will produce a flow of current I_{cc} in the external circuit given by the relation

$$I_{cc} = n_{eh} I_B = J_B E_0/e_{eh} \quad (7.2)$$

since each incident electron can produce n_{eh} carrier parirs. This signal, which we will call the charge-collected current I_{cc}, is seen to be substantially greater than the incident current I_B. In the scanning electron microscope (SEM), the current flowing around the external loop is measured and displayed as function of the beam position to produce the charge-collected, or EBIC, signal. Note that an important practical consideration is that the current we wish to measure is actually a short-circuit current. Hence external resistance load R_L must be of as low a value as possible or else the ohmic voltage drop across it, which is in the opposite sense to the potential at the depletion layer, will affect the signal collection process.

In the standard theory of Donolato (1978), two steps are necessary to compute I_{cc}: first, we must model the generation of carriers by the electron beam, and, second, we must model the transport and collection of the beam-generated carriers. Typically the generation of carrier pairs is described by a function $g(\mathbf{r})$, in units of $cm^{-3}s^{-1}$, modeled as a three-dimensional Gaussian Distribution (Fitting et al., 1977) or a sphere of uniform generation (Bresse, 1972), neither of which is a very realistic representation of the beam interaction with the solid. The transport of the minority component of the electron-hole pairs generated is a diffusion problem described by the equation:

$$D \nabla^2 p(\mathbf{r}) - \frac{1}{\tau} p(\mathbf{r}) + g(\mathbf{r}) = 0 \tag{7.3}$$

where $p(\mathbf{r})$ is the density of the beam-generated minority carriers and D and τ are, respectively, the minority carrier diffusion coefficient and lifetime. Given the necessary boundary conditions on $p(\mathbf{r})$ as defined by the geometry of the problem at hand, Eq. (7.3) can be solved to give $p(\mathbf{r})$ and the induced current I_{cc} can be found by integrating the normal gradient of p over the collection plane. While this description of the problem is rigorously correct, the drawback is that I_{cc} cannot be found without knowing $p(\mathbf{r})$, which, in turn, requires a knowledge of $g(\mathbf{r})$, and no realistic model of $g(\mathbf{r})$ yields an equation that is analytically tractable.

However, if we use the Monte Carlo procedures we have developed to describe the incident beam interaction with the semiconductor, then, at any instant along a trajectory, $g(\mathbf{r})$ is effectively a point source whose strength is equal to the number of electron-hole pairs generated in that segment of the trajectory, which for the kth trajectory step is equal (Akamatsu et al., 1981) to the energy given up ($E[k] - E[k + 1]$) divided by the energy to create one electron-hole pair e_{eh}. If the trajectory segment is at a distance s from the collecting junction, then the probability of $\psi(s)$ of these carriers diffusing to the collecting junction and producing a charge collected current is, (Wittry and Kyser, 1964), for a point source:

$$\psi(s) = \exp(-s/L) \tag{7.4}$$

where L is the minority carrier diffusion length (i.e., $L = \sqrt{D\tau}$).

It would thus seem that we could compute I_{cc} in a sequential manner by summing up the charge-collection contribution from each step of each of the trajectories that we simulate and averaging this to find effective charge collected per incident electron (Joy, 1986). Although this looks, at first sight, to be quite different from actually solving the diffusion problem of eq. (7.4) and then computing I_{cc}, these two approaches are, in fact, functionally equivalent. As shown by Possin and Kirkpatrick (1979), we can generalize the quantity $\psi(s)$ and define a quantity $\psi(\mathbf{r})$, which describes the probability that a minority carrier generated by an electron beam at $\mathbf{r}$ is collected and thus contributes to the charge-collected current I_{cc}. Donolato (1985, 1988) showed that for the geometry of Fig. 7.2C, $\psi(\mathbf{r})$ also satisfies Eq. (7.3) with the boundary condition $\psi(\mathbf{r}) = 1$ at the entrance surface ($z = 0$). In this case, the solution of Eq. (7.3) reduces to one dimension and has thc form

$$\psi(z) = \exp(-z/L) \tag{7.5}$$

That is, it is identical with the result of Eq. (7.4), and the charge-collected current I_{cc} is then the sum of the contribution of the elementary sources:

$$I_{cc} = \int_{\text{volume}} \psi(z)\, g(\mathbf{r})\, dV = \int_0^\infty \exp(-z/L)\, h(z)\, dz \tag{7.6}$$

where $h(z)$ is the depth distribution of the generation. Thus we can quite generally calculate the magnitude of I_{cc} for a given specimen geometry and beam interaction by first stepping through the Monte Carlo simulation to produce $h(z)$ and then using Eqs. (7.4) or (7.5) and (7.6) to find I_{cc} by simple summation.

7.3 Monte Carlo modeling of charge-collection microscopy

7.3.1 The generation function

Either the single or plural scattering models could be used as the basis for a simulation of charge-collection microscopy, but since the samples are invariably bulk, the plural scattering approach will be significantly faster and so will be illustrated here. The first task is to compute `eh`, the number of electron-hole pairs generated along each step of the trajectory. Working from the code in chap. 4, we can easily insert this as shown below. Since it is interesting to be able to view the actual spatial distribution of $g(\mathbf{r})$, we can associate each element of carrier production with the coordinates of the midpoint (xm, ym, zm) of the trajectory step. Since for normal incidence the distribution is going to be radially symmetrical about the beam axis, we can store this in an array $g(r, zz)$ where r, the radius from the axis, and zz, the depth, are both expressed in units of one-fiftieth of the Bethe range (i.e., the `step` length). Do not forget to add to the variables list the extended quantities `xm,ym,zm,eh,` the integers `r,zz,` and the array `g[0 . . . 50, 0 . . . 50]` of extended, and remember to initialize `g[r,zz]` before using it.

```
p_scatter; {find the scattering angles}
new_coord(step); {find where electron goes}

{program-specific code will go here}
  {******** and here it is ********}
```

```
        {eh the number of carriers generated is}
                eh:= (E[k]-E[k+1])*1000/3.6; {E[] is in keV}
    {assuming that the pair generation energy for silicon is 3.6 eV}
              xm:=(x+xn)/2;      {x coordinate midpoint of step}
               ym:=(y+yn)/2;      {y coordinate of midpoint}
              zm:=(z+zn)/2;      {z coordinate of midpoint}
        {get radius of midpoint from axis in units of range}
                r:=round(sqrt(xm*xm+ym*ym)/step);
                  zz:=round(zm/step);
        {trap any values that fall outside the specified array}
               if zz>=0 then
        {put this carrier contribution into the array at r,zz}
               g[r,zz]:=g[r,zz]+eh;
etc.
```

The array can be printed out at the end of the run to give the spatial distribution `g[r,zz]`. Because the data have been stored in units of a radial variable, the volume represented by successive values of *r* increases steadily, so, in order to make the values directly comparable, it is necessary to divide `g[r,zz]` by `(2r+1)`— i.e., the area between the `r`th and `(r+1)`th annuli. Figure 7.3 shows the carrier distribution calculated in this way for silicon. The profile has the familiar "teardrop" shape with a diameter of the order of the range, and with a maximum depth of about 0.75 of the range. It is clear, however, that the distribution of carrier generation is far from uniform, with nearly a quarter of the carriers being produced within a volume

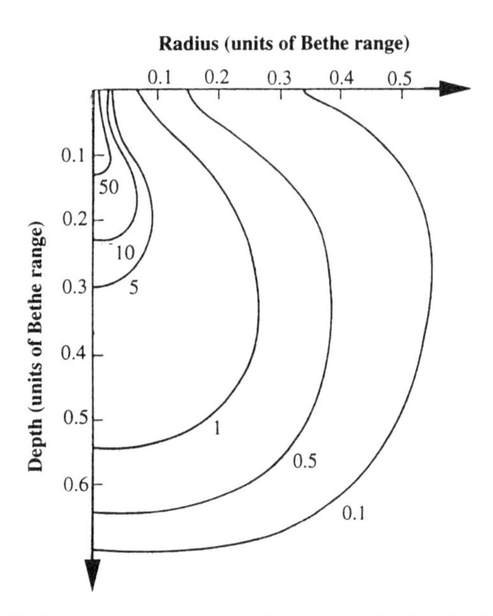

Figure 7.3. Isogeneration contours for electron-hole pairs in silicon.

that is only about 0.2 of the range in diameter. The isogeneration contours are in good agreement with published experimental data (e.g., Possin and Norton, 1975) confirming that the physical basis for the model is good.

7.3.2 The gain of a Schottky barrier

The specimen configuration most normally used for EBIC imaging is that of Fig. 7.2C, the Schottky barrier geometry; consequently, it is this system that we will use the Monte Carlo method to analyze, although other arrangements such as that using a p-n junction can as easily be studied by obvious modifications to the development below. The Schottky barrier consists of a thin layer, typically 100 to 300 Å, of a metal, such as gold for n-type silicon or titanium for p-type silicon, in intimate contact with the surface of the semiconductor. Although this film is thin, it may have a significant effect on the electron beam interaction with the solid, especially at low beam energies, and so must ultimately be included in the simulation. We assume that the Schottky produces a depletion region of depth *zd*. To a first approximation (Leamy, 1982)

$$zd = 0.53\ \Omega\ \sqrt{V_b} \qquad \mu\text{m} \tag{7.7}$$

where Ω is the resistivity (ohm.cm) and V_b is the barrier height, (e.g., for Si V_b = 0.7 V) plus any applied reverse bias. So for 1 ohm.cm resistivity silicon, with no external applied bias, *zd* is about 0.5 μm.

The measurable parameter of a Schottky diode in charge-collection mode is its gain *G*, which can be defined as:

$$G = \frac{I_{cc}}{I_b} \tag{7.8}$$

where I_{cc} is the measured short-circuit current collected for a given incident beam energy I_b. From the discussion above, we can see that G is of the order of n_{eh}, the number of electron-hole pairs generated per incident electron; but it will invariably be lower, because not all of these carriers will ultimately contribute to the signal. Let us consider a single step of a trajectory and calculate the incremental contribution to I_{cc}.

As before, the number of electron-hole pairs `eh` produced at this step is

$$\texttt{eh} = (E[k] - E[k + 1]/e_{eh}$$

and these we take to be generated at the depth $zm = (z + zn)/2$, the midpoint of the trajectory step. If $zm < zd$, then all of the carriers produced are separated, so the incremental charge collected `cc` is increased by `eh`. If $zm > zd$, then the carriers must diffuse back to the depleted region before they can be separated and collected.

The fraction of carriers collected `coll_frac` is, from Eq. (7.5),

$$\begin{aligned} \text{collfrac} &= \exp[-(zm - zd)/L] && \text{if } zm > zd \\ &= 1 && \text{if } zm <= zd \end{aligned} \tag{7.9}$$

where L is the minority carrier diffusion length.

In general, we want to know how the gain G depends on semiconductor parameters such as the depletion depth zd and diffusion length L. Because the collection of the carriers is totally independent of the generation process, we can conveniently do this by computing just a single set of trajectories but allowing zd and L to take a range of different values and calculating the gain in each case. If we define m values of the depletion depth $zd[]$ and n values of the diffusion length $L[]$, then we need a matrix $m*n$ in size for the current values *Icc[]*. The code fragment from above would then look like this:

```
p_scatter;          {find the scattering angles}
new_coord(step);    {find where electron goes}

{program-specific code will go here}
  {******** and here it is ********}
{eh the number of carriers generated is}
     eh:= (E[k]-E[k+1]*1000/3.6; {E[] is in keV}
{assuming that the pair-generation energy for silicon is 3.6
     eV}
     zm:=(z+zn)/2; {z coordiante of midpoint}

{compute the incremental current collected for each value of zd
          and L}
          for i:=1 to m do
            begin
             for j:=1 to n do
               begin
                 if zm<=zd[i] then {all the carriers are
                    collected}
                    Icc[i,j]:=Icc[i,j]+eh
                  else   {allow for diffusion}
                     Icc[i,j]:=Icc[i,j]+eh*exp(-(z-
                     zd[i])/L[j]);
               end;
            end;
          etc. . . .
```

Figure 7.4 shows the variation of the gain G computed in this way for silicon at 30 keV, as the depletion depth varies from 0.5 to 5 μm and as L varies from 0.1 to 5 μm. When the depletion depth `zd` or the diffusion length `L` are small compared to

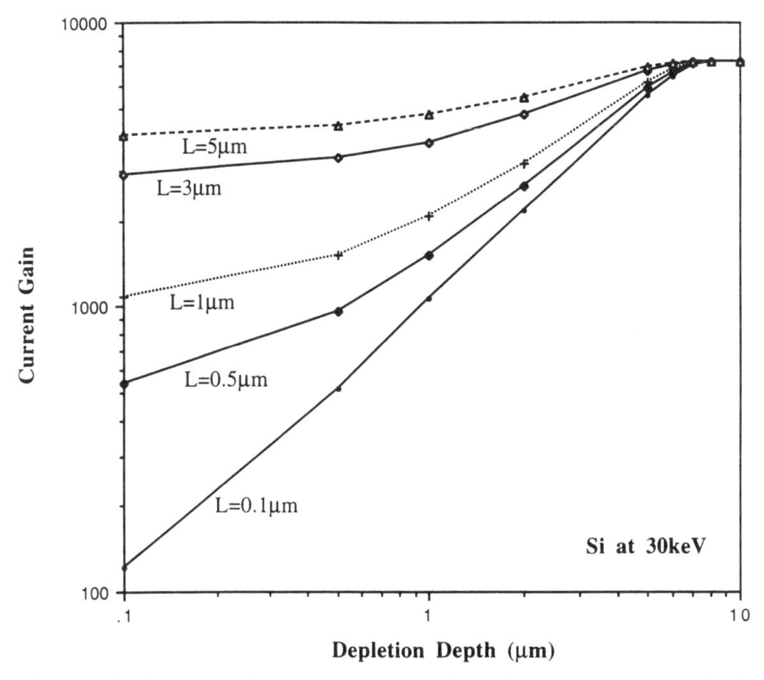

Figure 7.4. Variation of current gain in a silicon Schottky diode at 30 keV.

the electron range (i.e., about 10 μm), then the gain of the diode is a sensitive function of these parameters; but when either or both become comparable to the range, then the gain depends only on the beam energy. Note that the advantage of using the same set of trajectories to calculate the gain for each of the variables is not just that of saving computational time. Using the same set of random numbers for each piece of data ensures that the differences in the gain due to a change in the diffusion length or depletion depth are not masked because of the statistical variations between one Monte Carlo run and the next. An interesting exercise is to try the effect on these calculations of changing the angle of beam incidence (Jakubowicz, 1982; Joy, 1986). For a given depletion depth, diffusion length, and beam energy there is a range of angles over which the gain is constant. Since the angle of beam incidence can easily be varied—for example, by rocking the beam or mechanically tilting the sample—this provides a rapid experimental method for determining the diffusion length of a diode.

In general, we cannot ignore the effect on the incident electron beam of the Schottky barrier metal film. As the electrons pass through this, they lose energy and are scattered laterally. In addition, the effective backscattering coefficient of the specimen is changed. While these effects are small at moderate and high beam energies (15 keV and above), they are quite significant at lower beam energies because the metal film, though thin, is then a substantial fraction of the electron range. This situation can be modeled by slightly modifying the program BINARY discussed in Chap. 6. The physics is exactly the same as that described above, except that we only consider electron-hole pair generation in the substrate material (i.e., the semiconductor) and the distance that these have to diffuse before collection is measured to the bottom of the Schottky layer rather than to the surface. It is only necessary to add a few lines of code to BINARY to incorporate these calculations. That is, after the scattering calculation we write

```
{now allow the electron to be scattered}
                  p_scatter(i);
                  new_coord(i);
{test for electron position within the sample}
          if zn<=0 then
          begin
            bk_sct:=bk_sct+1;
             num:=num+1;
            goto back_scatter;
          end
          else
{compute the gain of the Schottky diode-energy of electron-
hole pair formation is e_eh-i.e., 3.6 eV for silicon etc.}
begin
      eh:=(E[i,k]-E[i,k+1])*1000./e_eh;    {number of eh pairs
      formed}
      if i=1 then eh:=0; {no carriers generated in metal film}
         if zn<= zd then {all of these carriers are collected}
              i_cc:=i_cc + eh
           else          {they have to diffuse back to depleted
                         region}
              i_cc:=i_cc + eh*exp(-(zn-zd)/L);
end;
          reset_next_step; {and so on . . .}
```

The only point to be careful of is to remember that, because this is a plural scattering simulation, the program cannot be expected to give accurate results above the energy for which the thickness of the barrier film becomes less than about one-tenth of the range (e.g., about 12 keV for a 300-Å gold film. Figure 7.5 shows the effect of including the Schottky barrier in attempting to compare experimental gain measurements with computed values, in this case for an indium phosphide (InP) diode. While the slope of the experimental data with energy is similar for both the case with no surface barrier and for the barrier thicknesses of 200, 300, and 400 Å, it is clear that the absolute gain values are quite different. The best match between the experimental data and the computed figures is for a barrier layer of 300 Å. If the gain can be measured over a sufficiently wide energy range (e.g., from a lower value where the gain is effectively zero because the beam is not penetrating the barrier to an upper value where the gain is at a maximum), then an iterative comparison

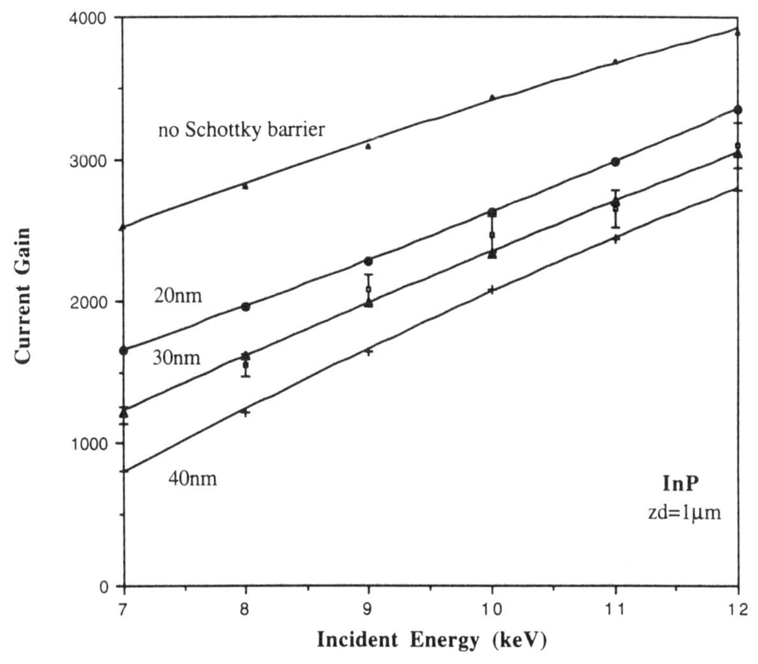

Figure 7.5. Experimental variation of gain with beam energy in InP and computed variation including thickness of Schottky layer.

between the Monte Carlo data and the experimental data allows both the parameters of the device (i.e., the diffusion length and depletion depth) and the thickness of the barrier to be determined.

7.3.3 Contrast from crystallographic defects

Contrast from electrically active defects in semiconductors was first observed in charge-collection images by Lander et al. (1963). Since that time the technique has been widely used (Leamy, 1982), not only because it allows defects to be imaged without the need to thin the material for transmission microscopy but also because, at the very low dislocation densities that are typical of today's semiconductors, EBIC imaging offers a much better chance of actually finding the defects in the first place. Contrast is seen from defects, because carrier recombination can occur at the defect, so reducing the number that contribute to the detected current. As shown by Donolato (1978) and using the notation shown in Fig. 7.6, a point defect at some depth D_d beneath the surface of a Schottky diode causes a differential signal change in the collected signal from a distribution of point sources $g(x, y, z)$ given by

$$dI = k. \int_{\text{vol}} g(x, y, z). \left[\frac{\exp\left(-\frac{r_1}{L}\right)}{r_1} - \frac{\exp\left(-\frac{r_2}{L}\right)}{r_2} \right] . \exp\left(-\frac{z}{L}\right) dv \qquad (7.10)$$

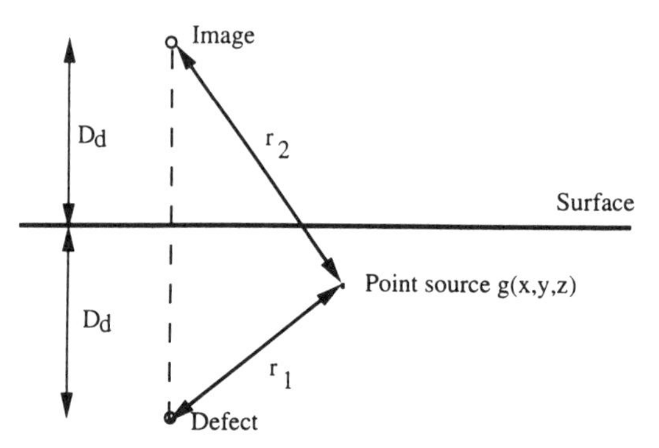

Figure 7.6. Definition of parameters used to model defects in a semiconductor.

where k is a measure of the recombination strength of the defect, r_1 and r_2 are defined as shown in Fig. 7.6, L is the minority carrier diffusion length, and the integral is over the whole excited volume. This derivation assumes that the Schottky barrier can be represented as a surface of infinite recombination velocity (i.e., any carrier reaching the surface contributes to the measured current).

As before, the Monte Carlo approach allows us to replace this integral by a summation in which the distribution of point sources $g(x, y, z)$ is represented by the carrier generation calculated for each step of the trajectory. For each step in the trajectory, the yield of carriers `eh` is found from Eq. (7.8). In the absence of a defect, the incremental current contributed to the imaging signal would then be `eh*coll_frac`, where `coll_frac` is given by Eq. (7.9) and depends on whether or not the center point of the step `zm` lies in the depleted region. In the presence of a defect, the actual collected signal increment will be `eh*coll_frac*defect_term`, where `defect_term` is from Eq. (7.10)

$$\text{defect\_term} = 1 - k.\left[\frac{\exp\left(-\frac{r_1}{L}\right)}{r_1} - \frac{\exp\left(-\frac{r_2}{L}\right)}{r_2}\right] \tag{7.11}$$

and represents the carriers lost to the defect and thus not contributing to the signal.

In general, line defects are of more interest than point defects in EBIC imaging. A line defect can be treated as a linear array of point defects of equal strength, so in order to model a dislocation it would be necessary to integrate Eq. (7.10) along the length of the defect, or in our discrete Monte Carlo model, to sum Eq. (7.11) along the dislocation line. If the dislocation is horizontal (i.e., parallel to the surface) and can be considered as infinite in length, then numerical integration (Joy, 1986) shows that the form of Eq. (7.11) remains unchanged except that `k`, representing the strength of the point defect, is now replaced by a term representing the strength per unit length of the line defect. This convenient simplification is possible because the

$1/r$ multiplying factors on each exponential in Eq. (7.11) greatly restrict the range over which the expression is significant. For other defect geometries—such as inclined faults or loops—it would be necessary to evaluate Eq. (7.11) for each trajectory step and for each point defining the defect line.

If the defect lies within the depleted region (i.e., $D_d < zd$), then the effect of the carrier drift, caused by the depletion field, on the image must also be considered. As shown by Leamy (1982), this can be done by comparing the transit time of carriers across the excited volume with the lifetime of carriers in the vicinity of the defect. The field at the defect E_D due to the sum of the barrier voltage V_b and the applied bias V_a is

$$E_D = \left(\frac{V_a + V_b}{zd}\right) \cdot \left(1 - \frac{D_d}{zd}\right) \tag{7.12}$$

where the depletion depth zd is given by Eq. (7.7). The effective strength per unit length of the defect is then reduced by a factor $E_0/(E_D + E_0)$, where E_0 is the field associated with the defect itself, typically of the order of 5.10^3 V/cm (Mil'stein et al., 1984).

As before, the basis of the simulation will be the BINARY program, and again only a minor modification to the original code is needed to include the computation of the effect of the defect:

```
{now allow the electron to be scattered}
              p_scatter(i);
              new_coord(i);
  {program-specific code will go here-so}
{first test for electron position within the sample}
          if i=2 then {the electron is in the silicon substrate}
             do_defect; {compute the current collected}
{now we can proceed as normal}
          if zn<=0 then {it's backscattered}
          begin
            bk_sct:=bk_sct+1;
             num:=num+1;
            goto back_scatter;
          end
          else
               reset_next_step; {and so on. . .}
```

As usual, all of the actual calculation is done in a procedure, here called `do_defect`. This, in turn, relies on some data that we have set up ahead of time. First, we must input at some convenient time the parameters that define the semiconductor and then calculate a few constants; for example:

```
{input the semiconductors' physical parameters}
    write('Resistivity of sample in ohm.cm');
      readln(resistivity);
    write('DC bias applied to Schottky (V)');
      readln(bias);
    write('depth of defect (μm)');
      readln(dd);
{now we can compute zd–the depletion depth of the material}
{assume the material is silicon and use Eq. (7.7)}
      zd:=0.53*sqrt(resistivity*(0.5+bias));
{we define the strength of the defect by giving it a "size" r0}
           r0=0.3; {nominal radius in microns}
     {now find ed, the field at the defect–Eq. (7.12)}
           if dd>zd then
                ed:=0 {defect is in neutral material}
           else
begin
       ed:=2*(0.5+bias)/thick; {thick is thickness of metal film}
       ed:=ed*1E4*(1-dd)/zd;
end;
 {and hence r_crit the effective size of the defect is}
       r_crit:=r0*sqrt(5E3/(5E3+ed));
 {assuming a nominal field of 500 v/cm from the defect itself}
```

Remember to add these variables to the `var` list at the start of the program—although, if you do not, the compiler will find them and warn you of their absence. Note that the constant k in Eq. (7.11), which represents the recombination strength of the dislocation, has the units of a length. So in the program, this has been called the "size" $r0$ of the defect, and been assigned a value of 0.3 μm. This value has been found to give contrast values that are in good agreement with those measured experimentally, and it is also consistent with the expected physical extent of the electrostatic field around a defect in silicon. For other materials or special conditions, such as a decorated defect, the value of $r0$ may have to be adjusted.

The presence or absence of the defect does not change the way in which the electron beam interacts with the solid. We can therefore again save time, and also improve our statistics, by simultaneously calculating the signal profile at many points across the dislocation for each trajectory step. If we place the dislocation so that it is parallel to the x axis and at $y = 0$, then for each trajectory step we can loop through the current calculation shifting the effective position of the beam by a series of values `pos[j]` and calculating the corresponding current `i_cc[j]`. Because the profile is symmetrical about the center of the defect, typically five to ten values in the array `pos[j]` are enough to produce a smooth profile. The chosen values for `pos[j]` can either be input by the user or preset in the program code itself. That is,

```
pos[0]:=0;
 pos[1]:=0.25; {all values are in micrometers from defect center}
 pos[2]:=0.5;
pos[3]:=1.0; {and so on}
```

A convenient rule of thumb is to set the maximum value of pos[j] to the sum of the electron range plus the diffusion length (i.e., typically in the range 5 to 15 μm for most materials and usable beam energies) and to choose intermediate values of pos[j] with approximately a constant ratio between them (e.g., 0, 0.25, 0.5, 1, 2, 4, 8, 16), so as to give the best definition to the central portion of the profile.

```
Procedure do_defect;
   {this calculates the current gain of the Schottky including the
          effect of the electrically active defect}
var r1,r2,del_I:real;    {local variables}
 begin                   {by computing number of e_h pairs generated}
     eh:=(E[i,k]-E[i,k+1])*1000/e_eh;
      zval:=(z+zn)/2; {midpoint of trajectory}
     yval:=(y+yn)/2; {midpoint of trajectory}

      for j:=0 to 10 do    {a loop shifting the defect position by
                             pos[j]}
         begin               {defect is at depth dd}
            r1:=(zval-dd)*(zval-dd)+(yval-pos[j])*(yval-pos[j]);
            {Fig. 7.6}
            r2:=r1+4*dd;    {by Pythagoras's theorem—see Fig. 7.6}
                        {now apply Eq. (7.11)}
                 del_I:=(exp(-r1/L)/r1 - exp(-r2/L)/r2);
                     del_I:=1-(r_crit*del_I);
                  if zval<=zd then {all of the carriers are
                             collected}
                             i_cc[j]:=i_cc[j]+eh*del_I
                  else    {they must first diffuse back}
                     i_cc[j]:=i_cc[j]+eh*exp(-(zval-zd)/L);
         end;          {of the loop across the defect}
 end;                   {of this procedure}
```

Because computations for ten or so different effective beam positions are made for each step of every trajectory, the program will run more slowly than normal. However, because the data from each of these points come from the same set of random numbers, the precision of the computation is high, and good statistics can be achieved for as few as 400 to 500 electrons. A plot of i_cc[j] against pos[j], possibly reflected about the position pos[0] = 0 so as to give a symmetrical profile, will now give the desired profile of the EBIC signal variation across the

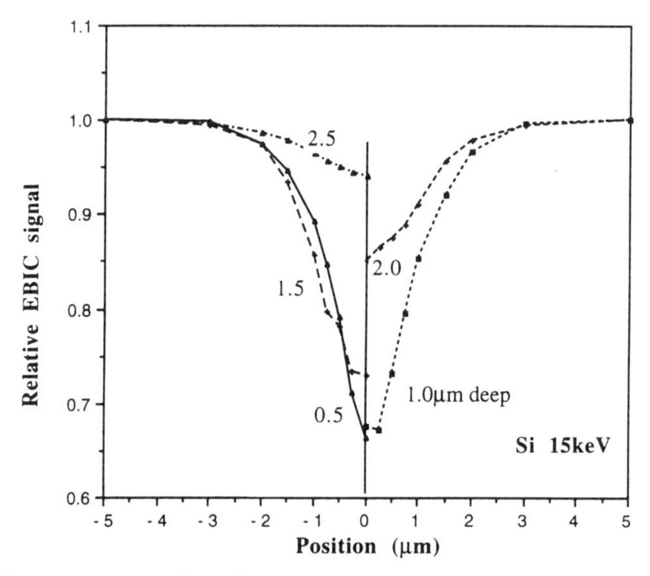

Figure 7.7. Computed EBIC profiles across a line defect at various depths in silicon.

defect. This can be done either by having the program output the actual `i_cc[j]` values and plotting these in some other program or by writing the necessary few lines of code to have the program plot its own output. In either case a great deal can be learned about the behavior of defect images in EBIC by using this program and trying the effects of various sets of semiconductor and defect parameters.

Figure 7.7 shows the predicted signal profiles arross a line defect at depths of 0.5, 1, 1.5, 2, and 2.5 μm in silicon. To avoid cluttering the figure, only half profiles are shown. With the physical parameters assumed here, a resistivity of 1000Ω.cm and zero applied bias, the depletion depth is [Eq. (7.7)] about 15 μm, which is large compared to both the electron range of 3 μm and the minority carrier diffusion length of 1 μm. For all depths, the profile has the same general V-shaped form, but the details of the profile vary from a narrow, high-contrast dip for a defect just 0.5 μm below the surface to a broad and rather shallow profile for a defect 2.5 μm below the surface. These predictions are in excellent agreement with experimental observations. See, for example, figure 33 in Leamy, 1982. It can also be noted that the contrast from a defect effectively vanishes if the depth of the defect exceeds the range of the incident electrons. This is because, as shown in Fig. 7.3, the majority of carriers are produced close to the surface, and it is evident from Eq. (7.11) that carriers generated far from the defect will contribute little contrast because the two exponential terms will be about equal and opposite. For good contrast to be observable from a defect, the carriers' generation maximum must be at or close to the defect. The depth at which a dislocation is situated can thus be estimated with good accuracy by noting the beam energy at which it first becomes visible and equating the electron range to the depth (Leamy et al., 1976; Joy et al., 1985).

Figure 7.8 shows the profiles predicted for the case where the defect is at a constant depth, here 0.5 μm, but the beam energy is varied. At 5 keV, where the beam range is also about 0.5 μm, the defect is just detectable, as would be expected from the discussion above. As the energy is raised, both ΔS, the signal change across the defect, and S, the total signal collected, increase, so the contrast $\Delta S/S$ also rises and reaches a shallow maximum at about 10 keV, at which condition the electron range, 1.5 μm, is about three times the defect depth. At still higher energies, the signal variation ΔS starts to fall while S slowly increases, so the contrast falls. Simulations for a wide variety of different defect parameters show that a contrast maximum can always be expected during a sweep of the incident beam energy and indicate that this maximum occurs when the defect is at a depth of between 0.3 and 0.4 times the beam range. The existence of such a contrast maximum with accelerating voltage was first predicted by Donalato (1978), who showed that, for a point defect and a uniform, spherical, generation volume, the maximum would occur at about 0.5 of the beam range. The knowledge that a dislocation at a given depth does not become visible until the beam range first reaches that depth, and that with increasing energy the contrast then goes through a maximum (at about twice the energy at which the defect first becomes visible) and then decays away, provides the

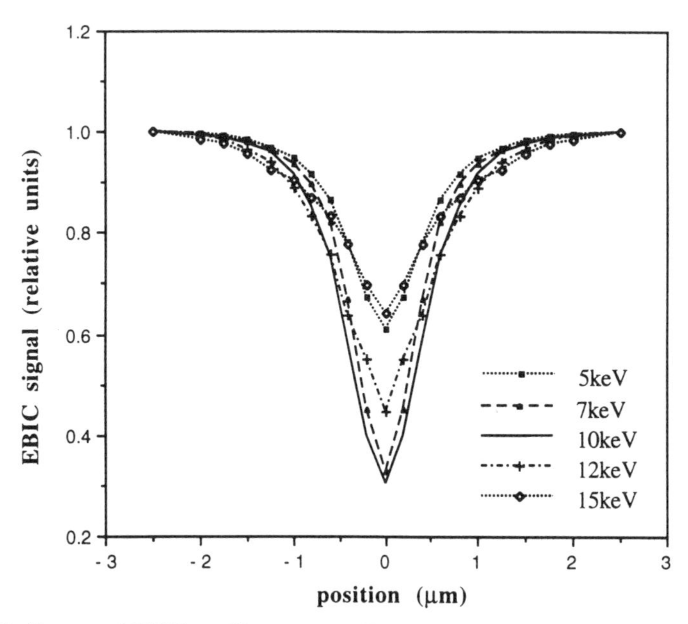

Figure 7.8. Computed EBIC profiles across a line defect in silicon as a function of beam energy.

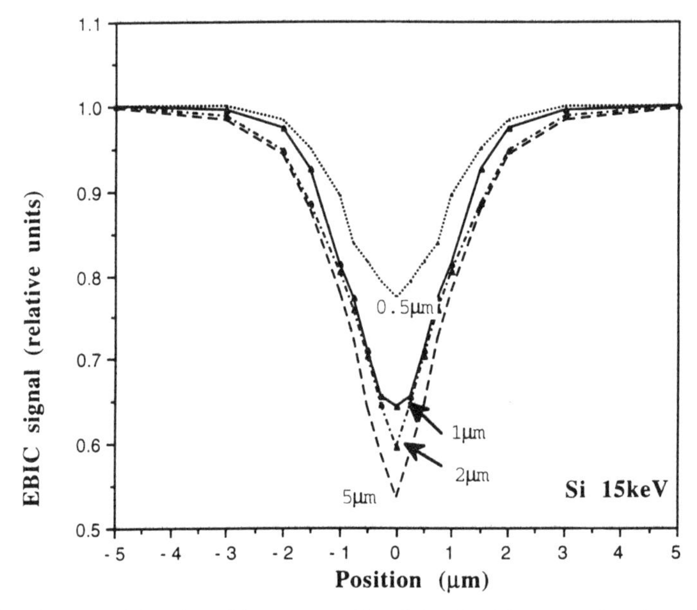

Figure 7.9. Computed EBIC profiles across a line defect in silicon as a function of diffusion length.

basis of a method for determining the three dimensional distribution of dislocations in a semiconductor (Joy et al., 1985).

It will also be expected that the minority carrier length L would have a significant effect on defect imaging in EBIC, since it is this parameter, together with the beam range, that sets a physical scale for the generation of the mobile carriers. Figure 7.9 shows how the signal profile across a dislocation, in this case 1 μm below the surface of 1000 Ω.cm silicon and viewed at 15 keV, is affected by the value of L. As the diffusion length increases, the shape of the profile remains essentially unchanged, but the apparent size of the feature increases. However, the full width at half maximum contrast of the defect image increases only by a factor of two times as L goes from 0.5 to 50 μm, and saturates if L is increased any further. In fact, measurements from simulations of this type show clearly that the spatial resolution of the EBIC image is almost independent of the diffusion length. The width of the defect image is determined by the beam energy (i.e., the lateral extent of the generation volume) and by the position of the defect relative to the surface. Consequently, even in materials with a long minority carrier diffusion length, high-resolution EBIC imaging is possible provided that the beam energy is reduced to minimize the interaction volume.

Although all of the examples given relate to EBIC imaging, the extension to C/L profiles is straightforward. All of the physics and program code associated with the generation of carriers and their interaction with defects remains unchanged, but all of the physics concerned with the subsequent collection of carriers is eliminated. In addition, since no depleted region is required, there is no practical reason to consider the presence of a Schottky barrier on the surface. Unless it is necessary to consider absorption of the C/L radiation as it leaves the specimen, the computation can be made straightforwardly by equating the resultant C/L signal with `i_cc`, the number of carrier pairs determined for each trajectory step. An example of the use of Monte Carlo simulations for both EBIC and C/L imaging is given in Czyzewski and Joy (1989).

The results discussed above demonstrate the power of the application of the Monte Carlo simulation methods to the analysis of a practical problem such as EBIC imaging. By carrying out a sequence of simulations, it is rapidly possible to determine what effect different experimental parameters will have on the observed profile. This, in turn, makes it possible to know how best to set up the microscope to achieve the desired result and allows a quantitative interpretation of the images generated to be made.

8
SECONDARY ELECTRONS AND IMAGING

8.1 Introduction

Secondary electrons (SE), discovered by Austin and Starke in 1902, are defined as being those electrons emitted from the specimen that have energies between 0 and 50 eV. Because of their low energy, secondary electrons are readily deflected and collected by the application of an electrostatic or magnetic field; as a consequence, the great majority of all scanning electron microscope (SEM) images have been taken using the SE signal. Secondaries can be generated through a variety of interactions in the specimen (Wolff, 1954; Seiler, 1984), a typical event being a knock-on collision in which the incident electron imparts some fraction of its energy to a free electron in the specimen. This is followed by a cascade process in which these secondaries diffuse through the solid, multiplying and losing energy as they travel, until they either sink back into the sea of conduction electrons or reach the surface with sufficient energy to emerge as true SE. For SE with energies of a few tens of electron volts, the inelastic mean free path (MFP) is small, in the range 10 to 40 Å, so each secondary typically travels only a short distance before sharing some of its energy in an inelastic event. However, for most materials, the inelastic MFP reaches a minimum at about 20 to 30 eV; for energies below that value, it increases rapidly, because there are no large cross-section inelastic scattering events through when energy can be transferred. The elastic MFP also falls with energy but becomes approximately constant at a few tens of angstroms for energies below about 30 eV. SE with energies below this value are therefore strongly elastically scattered even though inelastic scattering is insignificant. Consequently, as the cascade develops from some point below the surface of the irradiated specimen, only a finite fraction will actually reach the surface and escape to be collected. We can thus see that there is a region beneath the surface—the so-called SE escape depth, perhaps 50 to 150 Å in extent—beyond which no SE generated can reach the surface and escape.

The measure of secondary electron production is the SE yield δ, which is the total number of SE produced per incident electron. For all materials, δ varies with incident electron energy in the same manner as that shown in Fig. 8.1 for silver. The yield is low at high beam energies because most of the secondary production occurs too deep below the surface to escape, but it rises as the beam energy is reduced. Eventually a broad peak is reached, corresponding to the conditon where the inci-

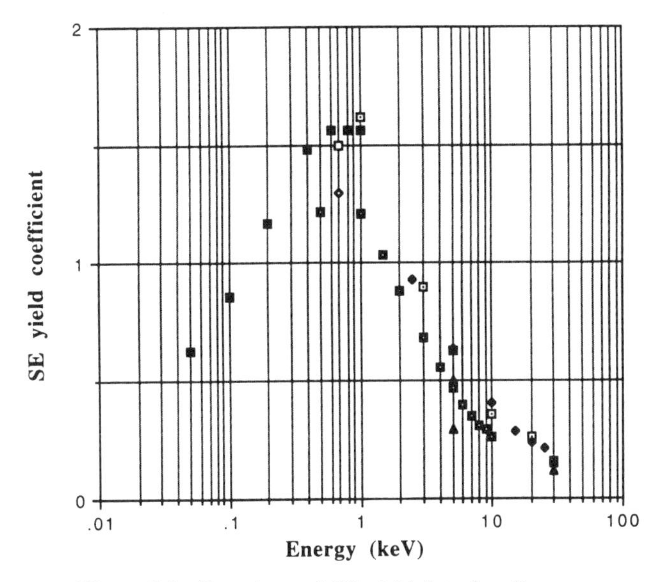

Figure 8.1. Experimental SE yield data for silver.

dent electron range is comparable with the SE escape distance. Typically this occurs for an energy of 1 to 3 keV. At still lower energies, the yield again falls because of the lower energy input from the incident beam. Any satisfactory simulation of SE production must be able to reproduce this yield curve for a given material and set of experimental conditions.

In general, there will be two occasions when SE generated by the beam can escape from the specimen surface: first as the incident electrons pass down through the escape zone and second as backscattered electrons again pass through this zone on their way back to the surface. While the SE produced in these two events are identical in nature, their utility for imaging is quite different, since the first group come from the beam impact point and are therefore capable of high spatial resolution while the second emerge from an area of the order of the incident electron range and are thus of low spatial resolution. It is thus convenient to denote them separately (Drescher et al., 1970). Those secondaries produced by incident electrons are called SE1, while those produced by the exiting backscattered electrons are called SE2. The ratio between these two components is also an important quantity to be able to calculate because it is a measure of the likely spatial resolution of the combined SE signal.

In this chapter we discuss three levels of approach to the problem of doing a Monte Carlo simulation of secondary electron production:

1. A complete first-principles simulation of all inelastic events leading to secondary production and including a model of the cascade multiplication of the SE and of their diffusion to the surface.

2. A simplified simulation in which one major mechanism for SE production is assumed to be dominant.
3. A parametric model of SE production and escape that can be added to a standard Monte Carlo model for incident electron trajectories.

The choice of which model to apply is determined by the information required. A first-principles simulation can provide detailed information about any of the parameters of SE production, e.g., energy, angle, and depth distribution of the SE in addition to the computed secondary yield δ. However, a substantial computing effort is required because of the complex nature of the physical models involved and, realistically, the application of this approach is limited to a very few elements and the most basic geometry. The simpler models typically calculate only SE yield, but they can provide this information in a short time even for structures of an arbitrary composition, size, and shape. These types of simulations are thus well suited for the computation of SE images, studies of charging, and microanalytical effects.

8.2 First principles—SE models

The secondary emission of SE has been the subject of a sustained body of theoretical research over a large number of years since its discovery (Austin and Starke, 1902). Important pioneer studies were performed by Bethe (1941) and Salow (1940), leading to the development of detailed phenomenological models by Baroody (1950), Jonker (1952), and Dekker (1958). Following Wolff (1954), who proposed the use of the Boltzmann transport equation to describe the process of SE generation, many authors, such as Cailler and Ganachaud (1972), Schou (1980, 1988) and Devooght et al. (1987) have used this approach. The Monte Carlo method has also been widely applied following the initial work of Koshikawa and Shimizu (1974). Construction of a Monte Carlo model for SE production involves three separate steps: determining the trajectory of the incident electron, computing the rate of SE generation along each portion of this trajectory, and finally calculating the fraction of SE that escape from the solid after the series of cascade processes.

The first part of this procedure is identical with what we have already considered at length, so we will not discuss it again here. Generally, it is sufficient to use a simple plural scattering method to model the incident electron trajectories. The second step requires us to take into account all possible creation processes for SE resulting from the interaction of primary electrons and backscattered electrons with free as well as with bound (i.e., core) electrons. In addition, the contribution to SE production from the volume plasmon decay should also be included. For each of these processes, we require an excitation cross section. The differential cross section for the production of SE from valence and d-shell electrons is (Luo et al., 1987)

$$d\sigma(E') = \frac{\pi e^4 dE'}{E\,E'^2} \tag{8.1}$$

where E is the energy of the incident electron and E' is the energy transferred to the secondary. The lowest allowed energy for an SE is chosen to be $E_F + \phi$ (where E_F is the Fermi energy and ϕ is the work function), so that the SE can cross the surface potential barrier and enter the vacuum state. Gryzinski's function (Gryzinski, 1965) is employed to describe the production of SE from the excitation of core electrons:

$$d\sigma(E') = \frac{\pi e^4}{EE'^2}\left(\frac{E}{E + E_j}\right)^{3/2}\left(1 - \frac{E'}{E}\right)^{E_j/(E_j+E')}$$

$$\times \left\{1 - \frac{E_j}{E} + \frac{4E_j}{3E'}\ln\left(2.7 + \left(\frac{E - E'}{E_j}\right)^{1/2}\right)\right\} dE' \qquad (8.2)$$

where E_j is the binding energy of the core electron.

For metals, the contribution to SE production by the decay of plasmons must also be considered. Using the expressions of Chung and Everhart (1977) gives an expression for the probability per unit distance of creating SE in the form:

$$\frac{d(1/\lambda)}{dE} = \lambda_{eff}^{-1}(E_0, \theta_1)\, D(E, h\omega_p, \Gamma_\upsilon) \qquad (8.3)$$

where

$$\lambda_{eff}(E_0,\theta_1) = \frac{2a_0E_0}{h\omega_p}\left[\ln\left(\frac{\theta_1^2 + \theta_E^2}{\theta_E^2}\right)\right]^{-1} \qquad (8.4)$$

and $D(E, h\omega_p, \Gamma_\upsilon)$, which describes plasmon decay by one-electron transitions, is

$$D(E,h\omega_p,\Gamma_\upsilon) = \sum_g n_g\, \mathbf{g}|\mathbf{W_g}|^2\left[\tan^{-1}\left(\frac{E - E_0^g - h\omega_p}{\Gamma_\upsilon\sqrt{2}}\right) - \tan^{-1}\left(\frac{E - E_F - h\omega_p}{\Gamma_\upsilon\sqrt{2}}\right)\right]$$

$$* \left\{\pi \sum_g n_g \mathbf{g}|\mathbf{W_g}|^2(E_F - E_0^g)\right\}^{-1} \qquad (8.5)$$

$$E_0^g = \frac{h\omega_p - h^2\mathbf{g}^2 - 4|\mathbf{W_g}|^2}{2\, h^2\mathbf{g}^2/m} \qquad (8.6)$$

where $h\omega_p$ is the plasmon energy, E_0 is the incident electron energy, a_0 is the Bohr radius, and W_g is the $\mathbf{g}^{th}$ Fourier coefficient of the lattice pseudopotential for the reciprocal lattice vector **g.**

It is usual (Bruining, 1954) to assume that the probability P_z of a secondary escaping from some depth z below the incident surface of the solid is given by a function of the kind

$$P_z = A \exp(-z/\lambda) \tag{8.7}$$

(where A is a constant of order unity) which is the "straight-line approximation" (Dwyer and Matthew, 1985) and implies that the emerging secondaries are unscattered and that any scattering of an SE produces absorption (i.e., only those SE that are not scattered between their point of generation and the surface can escape). In fact, these assumptions are not strictly valid (Chung and Everhart, 1977), but the error that they introduce is usually negligible. However, to accurately model the cascade, it is necessary to generalize Eq. (8.7). The probability of an SE arriving at the surface without any inelastic collision is

$$P_z = 0.5 \exp[-z/\lambda(E)\cos(\pi/4)] \tag{8.8}$$

because the average escape angle will be at 45° to the surface. $\lambda(E)$ is the inelastic mean free path for an SE of energy E, which for a metal is given (Seah and Dench, 1979) by the formula:

$$\lambda(E) = a\left\{\frac{538}{(E - E_F)^2}\right\} + 0.41a(E - E_F)^{1/2} \tag{8.9}$$

where E_F is the Fermi energy of the metal, and a is the thickness of a monolayer of the metal in nanometers. Typically $\lambda(E)$ is of the order of a few nanometers for energies in the sub-100 eV range.

We can now set up the cascade process. From Eq. (8.8), the probability P_z of an SE traveling from z to z' without any inelastic collisions is

$$P_{z'} = 0.5 \exp(-|z - z'|/\lambda(E)\cos(\pi/4)) \tag{8.9}$$

and similarly the probability of it traveling from z' to $z' + \Delta z'$ is

$$P_{z'+\Delta z'} = 0.5 \exp(-|z - (z' + \Delta z')|/\lambda(E)\cos(\pi/4)) \tag{8.10}$$

so the probability $\Delta P_{z'}$ that the SE has interacted with another electron (i.e., participated in the cascade) is

$$\Delta P_{z'} = P_{z'} - P_{z'+\Delta z'} = P_{z'}\,\Delta z'/\lambda(E)\cos(\pi/4) \tag{8.11}$$

Some SE thus travel to the surface at a rate governed by the exponential decay law while others take part in the cascade process and produce new SE of lower energies. It is convenient to make a distinction between secondaries of different energy ranges in the cascade. Below 100 eV, the scattering of the SE can be assumed to be spherically symmetrical (Koshikawa and Shimizu, 1974). If E' is the energy of an SE after scattering, then

$$E' = E\sqrt{\text{RND}} \tag{8.12}$$

where E is the energy before scattering and RND is the usual equidistributed random number. Thus for each SE participating in the cascade, two SE appear after a collision with energies E' and E'', where

$$E'' = E(1 - \sqrt{\text{RND}}) \tag{8.13}$$

Above 100 eV, the scattering of the SE is given by the usual Rutherford relations, and Eqs. (8.1) and (8.2) are used to calculate the rate at which new SE are generated by core electron ionizations and from valence and d electrons. The calculation of the cascade process is carried out up to a sufficiently large depth from the incident surface to ensure that all contributions are accounted for.

The final step in the analysis is to calculate what fraction of the SE reaching the surface actually pass the energy barrier and reach the vacuum. The elastic scattering of the SE can be assumed to be isotropic, because at low energies the elastic MFP is only of the order of atomic dimensions (Samoto and Shimizu, 1983), so all directions of motion are equally probable for the internal SE. In order for an SE of energy E to escape, E must be greater than $E_F + \phi$, where E_F is the Fermi energy and ϕ is the work function. The maximum angle α at which the SE can approach the surface is determined by taking the normal component of momentum $P\cos\alpha$ equal to the value

$$P_c = \sqrt{2m}(E_F + \phi) \tag{8.14}$$

Thus the escape probability $p(E)$ for an SE of energy E at the surface is

$$p(E) = \frac{\int_0^{\alpha} \sin\alpha \, d\alpha \int_0^{2\pi} d\phi}{2\pi} = 1 - \cos\alpha = 1 - \frac{P_c}{P} = 1 - \sqrt{\frac{E_F + \phi}{E}} \tag{8.15}$$

Putting all of the pieces described above together now gives us a complete Monte Carlo description of the generation, multiplication, diffusion, and escape of the secondary electrons from a solid. The actual code will not be reproduced here because, of necessity, it is lengthy and complex. When applied to the problem of computing SE properties from selected metals—usually Al, Cu, Ag, and Au—detailed agreement with experimental data is obtained. For example, as shown in Fig. 8.2, (adapted from Luo and Joy, 1990), the computed normalized secondary energy distribution $N(E)$ is in good agreement with the available experimental data. The agreement between computed and measured SE yields is also quite good, although—as was the case for backscattered yield data—the spread between different sets of experimental results is often very large, making any quantitative assess-

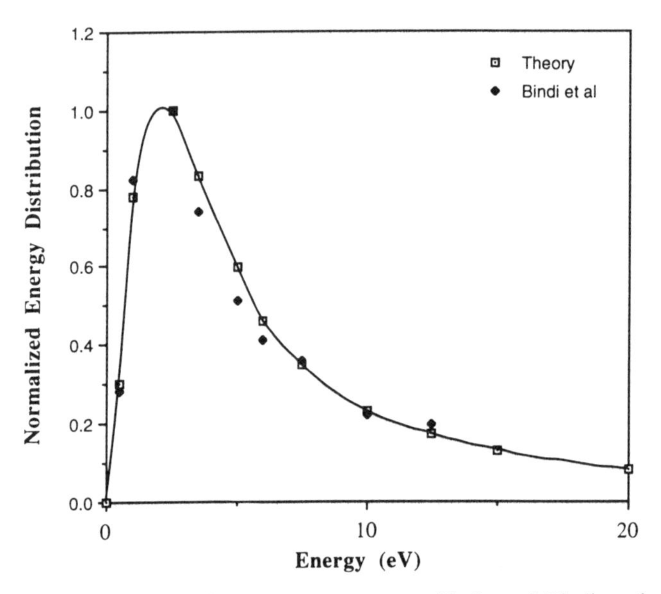

Figure 8.2. Comparison of SE energy spectrum with data of Bindi et al. (1980).

ment of accuracy a difficult matter (Joy, 1993). It is usual to assume that the angular distribution of emitted secondaries follows a cosine law (Jonker, 1952), a result that follows directly from the isotropic nature of elastic scattering within the specimen (Luo and Joy, 1990). However, the detailed simulation shows that when the inelastic scattering that occurs within the cascade is taken into account, there are deviations from the cosine rule especially for the higher-energy SE. Figure 8.3 shows how this deviation appears. Although the effect is real, the fractional number of such high-energy SE is small, and so—for most practical purposes—the cosine rule can be taken to apply. The model also calculates how the yield of SE varies as the thickness of the target is increased. The SE1 component of the SE emission, that produced by the forward-traveling incident electrons, typically reaches its maximum value for a thickness of only a few nanometers. As the thickness is increased beyond that value, however, the number of backscattered electrons continues to rise, and so the yield of SE2 secondaries increases and the total yield δ rises. Figure 8.4 shows experimental data for copper, illustrating this effect.

Since

$$\delta = SE1 + SE2 \tag{8.16}$$

we can write

$$\delta = SE1(1 + \beta\,\eta) \tag{8.17}$$

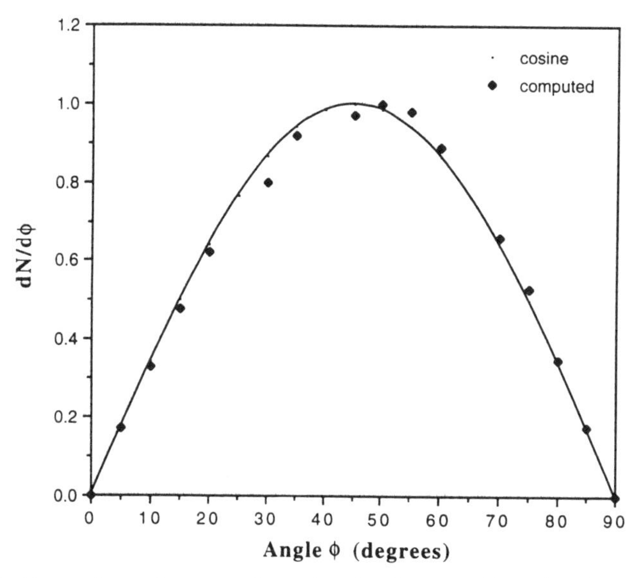

Figure 8.3. Comparison of computed angular distribution with cosine law, Al at 2 keV.

where η is the backscattering coefficient for the target and β is a factor that represent the relative efficiency of backscattered electrons at generating SE. If SE1 is constant, as is the case for any target thicker than a few nanometers, then the variation of δ with thickness provides an experimental way to measure the value of β, since we know that η varies linearly with thickness (see Chap. 6). This indirect approach must be used because we cannot physically distinguish SE1 and SE2 electrons, since their properties are identical. However, in a calculation, it is a simple matter to compute the SE1 and SE2 yield separately and hence to obtain a value for β. The computation shows that β is typically in the range 3 to 6, the value being somewhat higher at low energies ($\approx$1 keV) and for materials with higher atomic numbers. This sort of value is in fair agreement with the available experimental values (e.g., Bronstein and Fraiman, 1961) and indicates that in a typical metal or semiconductor for which $\eta \approx 0.3$, less than 50% of the total SE signal is being produced by the incident beam.

The use of this model with other than a few familiar metals is difficult, primarily because many of the required parameters in the model are either unknown or known only imperfectly. For nonmetallic elements such as semiconductors and dielectrics, significant modifications to the model would be required to make it physically realistic. Work in this area is still proceeding. Thus, while this method is of theoretical importance in studies of the phenomena associated with SE emission, in the context of this book the model is of limited value because it cannot be applied to the real-world situation in an SEM. For this kind of problem, simpler, less detailed, but more pliable, models must be used.

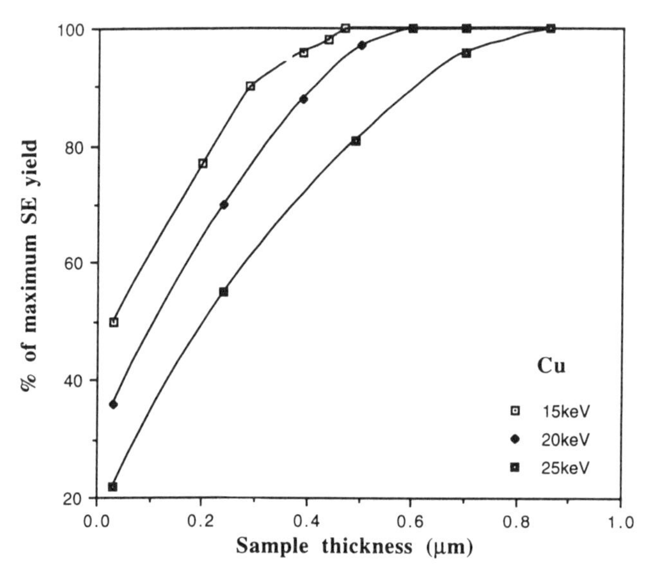

Figure 8.4. Variation of SE yield with sample thickness.

8.3 The fast secondary model

We can simplify the complexity of the previous method by assuming that only a single mechanism for the production of secondary electrons need be considered. As originally suggested by Murata et al. (1981), an appropriate assumption is that the SE are produced by a knock-on collision in which an incident electron interacts with a free electron. The required differential cross sections for this type of interaction are poorly known; however, this type of interaction can be treated classically by considering it as a coulomb interaction. In this case, the cross section per electron is (Evans, 1955)

$$\frac{d\sigma}{d\Omega} = \frac{\pi e^4}{E^2}\left(\frac{1}{\Omega^2} + \frac{1}{(1-\Omega)^2}\right) \tag{8.18}$$

where E is the incident electron energy and ΩE is the energy of the secondary produced. Since this interaction involves energy transfers of ΩE and $(1 - \Omega)E$ and since, in the final state, the incident electron will no longer be distinguishable, the two cross sections must be added. For convenience, we can define the electron with the highest energy as the primary; thus Ω is restricted, so that $0 < \Omega < 0.5$. The inelastic scattering (Fig. 8.5) causes a deflection α of the primary electron in the laboratory frame of reference given as:

$$\sin^2\alpha = \frac{2\Omega}{(2 + t - t\Omega)} \tag{8.19}$$

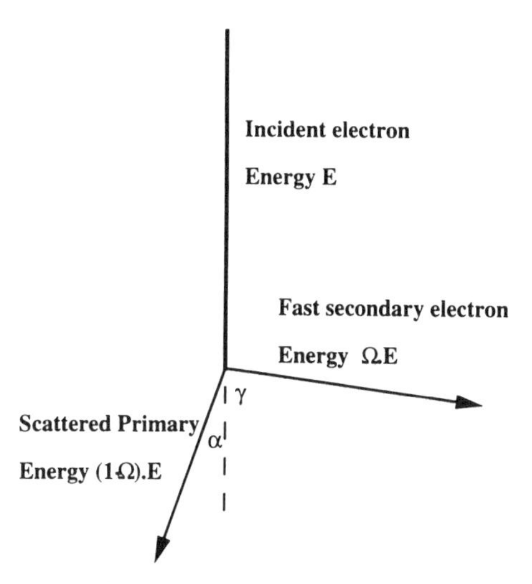

Figure 8.5. Geometry of scattering event to produce a fast secondary electron.

where t is the kinetic energy of the electron in units of its rest mass (511 keV). For a small energy transfer, say 500 eV for a 20-keV primary, α is about 1°; which is thus of the same general order of magnitude as the average elastic scattering angle. However, the secondary electron leaves the impact point at an angle γ given by

$$\sin^2\gamma = \frac{2(1 - \Omega)}{(2 + t\Omega)} \tag{8.20}$$

so that the 500-eV SE leaves at an angle of about 80° to the initial direction of travel of the primary electron.

Other cross sections for SE production by knock-on have been given, for example by Mott (1930), Möller (1931), and Gryzinski (1965). These models differ in the assumptions that they make and in their functional form, but the predicted values of the cross sections under equivalent conditions differ relatively little provided that the energy of the SE is sufficiently high for it to be considered "free." Since there are no experimental data clearly supporting one cross section over another, we will use the Evans model [Eq. (8.18) and following], since the values predicted by this fall in the middle of the range of numerical values spanned by the other formulations.

8.3.1 Constructing a fast secondary Monte Carlo model

The fast SE (FSE) model that we are developing here is a double Monte Carlo simulation, since we track an incident electron, as usual, but if a secondary electron is generated, we freeze the position of the primary, track the secondary until it leaves the specimen or loses it energy, and then resume tracking the incident electron again. The model that we will use is the single scattering model of Chap. 3 because we have to investigate each electron interaction individually to see if it is an elastic or an inelastic event. We will use the program to plot the trajectories of either the incident or the FSE and to calculate the yield of SE from the specimen. Below is the key portion of the original listing of the single scattering program of Chap. 3, but now modified to allow for FSE generation. The changes and additions to the original code are shown in boldface.

```
{***********************************************
 *              the Monte Carlo loop            *
 ***********************************************}

                    zero_counters;

                 while num < traj_num do
begin
                    reset_coordinates;

               {allow initial entrance of electron}
                  step:=-lambda(s_en)*ln(random);
                     zn:=step;

              if zn>thick then {this one is transmitted}
                begin
                     straight_through;
                     goto exit;
                end
              else   {plot this position and reset coordinates}
                begin
                     xyplot(0,0,0,zn);
                     y:=0;
                     z:=zn;
                end;

       {now start the single scattering loop}
repeat

{***********************************************
 *              is an FSE produced?             *
 ***********************************************}

  Test_for_FSE:=RND;
       if Test_for_FSE>PEL then {we have generated an FSE}
```

```
begin
    Track_the_FSE; {follow till it escapes or dies}
      goto reentry;    {then go back to main program}
end;
    {otherwise scatter the primary electron in the normal way}
            s_scatter(s_en);

    reentry: {FSE program rejoins main loop}

      step:=-lambda(s_en)*ln(random);

          new_coord(step);

    {problem-specific code will go here}

          {decide what happens next}

    etc.
```

It can be seen that, at least at this level of detail, the changes involved are minimal. The initial entry of the electron proceeds as before, but now—instead of just allowing the electron to be scattered elastically—we test to see whether or not an inelastic scattering event, resulting in the production of a fast SE, has occurred. This is done by seeing if a random number `Test_for_FSE` is greater or less than a variable `PEL`, which is the ratio of the elastic cross section to the total scattering cross section (i.e., including the inelastic effect).

The elastic scattering undergone by the incident electron is represented as usual by the Rutherford cross section [Eq. (3.2)]. This cross section σ_E defines a mean free path λ_{el} given by the formula

$$\lambda_{el} = \frac{A}{N_a \rho \sigma_E} \tag{3.3}$$

which represents the average distance that an electron will travel between successive elastic scattering events. Similarly we can define an inelastic MFP λ_{in} by integrating Eq. (8.18). However, we note from Eq. (8.18) that the cross section becomes infinite at $\Omega = 0$, so a lower limit Ω_c must be chosen as a cutoff. Choosing a finite value of Ω_c not only avoids the problem of an infinite cross section but also has the effect of removing from consideration very low energy secondaries which, because of their small MFP, require substantial computer time during the simulation without contributing much to the final result. It is found (Murata et al., 1981) that the choice of Ω_c does not sensitively affect the results produced, since lowering Ω_c produces more secondaries but each of a lower average energy. So, for convenience,

Table 8.1 Values computed for 20-keV beam incidence

Element	λ elastic	λ inelastic
Carbon	338 Å	927 Å
Aluminum	289 Å	1020 Å
Silicon	240 Å	871 Å
Copper	54 Å	250 Å
Silver	36 Å	198 Å

a value of 0.01 will be used here. Equation (8.18) can then be integrated over the range $\Omega_c < \Omega < 0.5$ to give the total inelastic cross section σ_{in} and hence

$$\lambda_{in} = \frac{A}{N_a Z \rho \sigma_{in}} \tag{8.21}$$

since the cross section is per electron. λ_{in} is the average distance between successive inelastic scattering events. Whereas λ_{el} is of the order of 100 Å for most materials at 20 keV, λ_{in} is about 0.1 μm at the same energy, so the probability of an inelastic collision is small compared to that for an elastic event. Table 8.1 gives some typical values for λ_{el} and λ_{in}. We can define a total scattering MFP—the average distance between scattering events of either type—from the relation

$$\frac{1}{\lambda_T} = \frac{1}{\lambda_{el}} + \frac{1}{\lambda_{in}} \tag{8.22}$$

`PEL` is the ratio λ_T/λ_{el} and represents the probability that a given scattering event will be elastic. Since $\lambda_{in} \gg \lambda_{el}$, then `PEL` is close to unity and only a small fraction of scattering events will be inelastic. The type of event that occurs is determined by choosing a random number `Test_for_FSE` and seeing if this is greater than `PEL`. If it is not, then the scattering event was elastic, no FSE was generated, and the program proceeds as normal. If, however, `Text_for_FSE > PEL`, then an FSE has been produced and the routine `Track_the_FSE` is called to track the FSE while the tracking of the incident electron is suspended. It is convenient to be able to turn FSE production on and off as required. This can be done in the setup of the program:

```
GoToXY(some suitable screen coordinates);
    Write('Include FSE (y/n)?');
     if yes then
       FSE_on:=true
      else
       FSE_on:=false;
```

This sets a Boolean variable FSE_on to true if FSE production is required, and to false if FSE are not needed. If FSE_on is true, then PEL is calculated as described above; but if FSE_on is set false, then PEL is set equal to unity, $\lambda_T = \lambda_{el}$, and the program behaves in exactly the same way as the original single scattering model.

The calculation of PEL is carried out in the function lambda(s_en), which, although it has the same name as before, is now modified to incorporate the inelastic scattering.

```
Function lambda(energy:real):real;
   {this now computes both the elastic and inelastic MFP}

  var al,ak,sg, mfp1,mfp2,mfp3:real;
        begin   {the function}
            if energy<e_min then {don't allow anything below cutoff
                     energy}
                     energy:=e_min;
                   al:=al_a/energy;
                       ak:=al*(1.+al);

           {this gives the elastic cross secton sg in cm² as}
              sg:=sg_a/(energy*energy*ak);
           {and hence the elastic mfp in Å is}
              mfp1:=lam_a/sg;
           {from Eq. (8.18) the MFP for FSE in Å is}
              mfp2:=at_wht*energy*energy*2.55/(density*at_num);
            {so the total MFP from Eq. (8.22) is}
                   mfp3:=(1/mfp1)+(1/mfp2);
                   mfp3:=1/mfp3;

            {make provision to switch FSE generation off}
               if FSE_on then {compute the ratio constant
                        elastic/total}
                        PEL:=mfp3/mfp1
                           else
                             begin
                                PEL:=1; {no FSEs will be produced}
                                  mfp3:=mfp1; {and λT = λel}
                             end;

               {in either case the value returned is}
                                lambda:=mfp3;
  end;
```

The final part of the operation is to track the FSE that are produced. This involves several steps that can be displayed schematically as

```
Calculate the energy of the FSE
  Store the coordinates of the incident electron at the point where
    scattering occurred
    Find the scattering angles of the FSE and its MFP
     Find the end point of its first trajectory step
    Follow FSE through standard single scattering Monte Carlo loop
      until FSE leaves the specimen or falls below the energy cutoff
      Calculate scattering angle of incident electron from inelastic
      event
    Find energy of incident electron after inelastic event
Return to main program
```

The procedure `Track_the_FSE` and an associated function `FSEmfp` carry out these tasks.

```
Function FSEmfp(Energy:real):real;
   {computes elastic MFP of FSE during its subsequent scattering}

var QK,QL,QG:real;

  begin
          QL:=power(at_num,0.67)*3.4E-3/energy;
            QK:=QL*(1+QL);
            QG:=(at_num*at_num)*9842.7/(energy*energy*QK);
          FSEmfp:=(at_wht*1E8)/(density*QG);
  end;

Procedure Track_the_FSE;
   {generates an FSE and then tracks it. This procedure is a complete
    single scattering MC loop. Note that all variables are local.}

  label set_up_reentry,FSEloop;

  var
     FSEnergy,eps,sp,cp,ga,FSE_step:real;
     S1,S2,S3,S4,S5,S6,S7:real;
     v1,v2,v3,v4,an_m,=an_n:real;
     deltaE,al,escape:real;

  begin
   {increment the counter for FSE production}
                        FSE_count:=FSE_count +1;

               {get the energy of the FSE that is produced}
                     eps:=1/(1000-998*RND);
                     FSEnergy:=eps*s_en;
```

```
{we now store the coordinates of the primary electron to reuse
        them later}
        S1:=x;
            S2:=y;
               S3:=z;
                  S4:=cx;
               S5:=cy;
            S6:=cz;
        S7:=s_en-s_en*eps: {incident energy loss = FSE energy}

    {see if the FSE exceeds the minimum energy that we want to
      consider}
      if FSEnergy<e_min then {its not worth tracking}
        begin
          escpae:=750*power(FSEnergy,1.66)/density;{estimate
          range}
          se_yield:=se_yield+0.5*exp(-z/escape); {contribution
            to yield}
            if FSE_on then xyplot (y'z'y+1'z+1); {plot a dot}
            goto set_up_reentry; {and out of here}
        end;

{otherwise get the initial scattering angles for the FSE that are
                  produced}
                  sp:=2*(1-eps)/(2+eps*FSEnergy/511.0);
                            cp:=sqrt(1-sp);
                           sp:=sqrt(sp);

         FSEloop: {single scattering model loops round here}

                     {find the FSE step length}
                   FSE_step:=-FSEmfp(FSEnergy)*ln(RND);
                       ga:=two_pi*RND;
     {find out where the FSE has gone using usual formula}

   if cz=0 then cz:=0.0001; {avoid division by zero}
                   an_m:=(-cx/cz);
                   an_n:=1.0/sqrt(1.+(an_m*an_m));

      {save time by getting all the transcendentals first}
                   v1:=an_n*sp;
                    v2:=an_n*an_m*sp;
                    v3:=cos(ga);
                   v4:=sin(ga);

            {find the new direction cosines}
                ca:=(cx*cp) + (v1*v3) + (cy*v2*v4);
                 cb:=(cy*cp) + (v4*(cz*v1 - cx*v2));
                cc:=(cz*cp) + (v2*v3) - (cy*v1*v4);
```

```
            {get the new coordinates}
                xn:=x + FSE_step*ca;
                 yn:=y + FSE_step*cb;
                zn:=z + FSE_step*c

               if zn>thick then {this one is transmitted}
                   begin
                 if FSE_on then xyplot (y,z,yn,999);
                    goto set_up_reentry; {get out of this
                                   function}
                                   end;
               if zn<=0 then       {its an emitted SE}
                   begin
                    if FSE_on then xyplot (y,z,yn,99);
                      se_yield:=se_yield+1;
                       goto set_up_reentry; {get out of this
                       function}
                   end;
   if FSE_on then xyplot (y,z,yn,zn); {plot the other case}

           {find the energy loss of FSE on this step}
               deltaE:=FSE_step*stop_pwr(FSEnergy)*density*1E-8;
            {so current FSE energy is}
                FSEnergy:=FSEnergy-deltaE;
      if FSEnergy<=e_min then {stop tracking it}
         begin {estimate escape probability to surface}
            se_yield:=se_yield+0.5*exp(-(z+zn)/(2*se_escape));
               goto set_up_reentry;
         end
               else {go round again}
                  x:=xn;
                   y:=yn;
                    z:=zn;
                    cx:=ca;
                   cy:=cb;
                  cz:=cc;

             {scatter the FSE}
              al:=al_a/FSEnergy;
                cp:=1-((2*al*RND)/(1+al-RND));
                 sp:=sqrt(1-cp*cp);
              goto FSEloop; {round again}

set_up_reentry;    {otherwise we exit the loop back to main program}
       {reset all variables to their entry values}

                  x:=S1;
                   y:=S2
```

```
                z:=S3;
                cx:=S4;
                cy:=S5;
               cz:=S6;
             s_en:=S7;

   {get scattering angles for the primary electron as a result of
inelastic event}
          sp:=(eps+eps)/(2+(s_en/511)-(s_en*eps/511));
             cp:=sqrt(1-sp);
             sp:=sqrt(sp);
          ga:=two_pi*RND;

   {and now we return back to the main program}

 end;
```

The first job of the procedure is the compute the energy of the FSE produced. This is done by solving for Ω [e.g., see Eq. (1.1)] the equation

$$\mathrm{RND} = \frac{\int_{\Omega_c}^{\Omega} \left(\frac{d\sigma}{d\Omega}\right) d\Omega}{\int_{\Omega_c}^{0.5} \left(\frac{d\sigma}{d\Omega}\right) d\Omega} \tag{8.23}$$

where $(d\sigma/d\Omega)$ is given by Eq. (8.18) and Ω_c is the cutoff discussed above. Evaluating this gives

$$\Omega = 1/(1000\text{-}998^*\mathrm{RND}) \tag{8.24}$$

which can be seen to yield properly weighted values of Ω between 0.5 and 0.001 for $0 \leq \mathrm{RND} \leq 1.0$. The actual energy of the FSE produced is then

$$\mathrm{FSEnergy} = \Omega^*\text{energy of incident electron}$$

The coordinates describing the position, direction of motion, and energy of the incident electron are next stored in the local variables, `S1, S2 . . . S7`. The quantities `S1` through `S6` hold `x,y,z,cx,cy,` and `cz` respectively. `S7` is set equal to the energy of the incident electron after the inelastic collision that equals $(1 - \Omega)^*$ incident energy, since th total energy is conserved. In order to save time, we first check the energy of the FSE found in Eq. (8.24). If this is below an arbitrarily chosen cutoff energy of 500 eV (not to be confused with the generation cutoff Ω_c), then the position where the FSE was generated is plotted. Because secondaries of this low an energy do not travel more than a few tens of angstroms in the material, there is no point in tracking them. However, they might contribute to the overall yield, since there is a finite probability that the FSE could reach the entrance surface and escape to be collected as a secondary electron. The chance of this occurring is estimated by finding the range `escape` for the electron at the energy `FSEnergy`

```
{estimate the FSE escape distance from range equation}
escape:=750.*power(e_min,1.66)/density; {in Å}
```

and then applying Eq. (8.7) to get the probability $p(z)$ of the SE traveling the distance z back to the surface:

$$p(z) = 0.5 \exp(-z/\text{escape}) \tag{8.25}$$

The factor 0.5 accounts for the fact that the SE can travel in any direction, so on average only half the secondaries will travel toward the entrance surface. The `se_yield` counter is then incremented by this fractional amount $p(z)$ and the function returns to the main program. This is clearly not a rigorous approach to accounting for the behavior of the FSE below the cutoff energy, but it is a plausible first approximation.

Otherwise, if the energy of the FSE is about 500 eV, then the angle `sp` at which the FSE leaves its generation point relative to the primary electron's direction of travel is computed from Eq. (8.20), the azimuthal direction `ga` is found from a random number call, and the step length for this event is found by computing the elastic MFP for the FSE at the energy `FSEnergy`, using the function `FSEmfp` from the expressions developed in Chap. 3. Note that we do not consider the possibility of the FSE in its turn having an inelastic collision and producing a tertiary electron, although this could be done with little additional programming if required.

The new coordinates of the FSE can now be found again, using our usual formulas. It might, at first sight, seem wasteful to include these equations in this `Track_the_FSE` procedure rather than to use the procedure already included in the main program, which does the same job. The reason this is done is that the normal procedure employed, `newcoords`, uses global variables (`x,y,z,cx,cy,cz, sp,cp,ga` etc.) to do its calculations (for a definition of global and local variables, see Chap. 2). If we want to use this same procedure to handle the FSE, then we would have to store the global variables (since these refer to the incident electron), substitute the appropriate quantities for the FSE, perform the calculation, and then swop back the original parameters. It is less confusing to use a purely local version of this and the other calculations required (e.g., the elastic MFP computation for the FSE), and it also removes a likely source of errors and bugs in the program code.

The new coordinates of the FSE are now tested. If the FSE has left the bottom surface of the specimen, then it has been transmitted, so the tracking is terminated, the coordinates of the incident electron are picked up from their temporary storage, and the function exits back to the main program. If the FSE has left through the entrance surface, then the counter `se_yield` is incremented by 1, since a detectable secondary electron will have been produced, and the function then exits back to the main program. If the FSE is still in the specimen, then the energy of the FSE is corrected for the amount that it has lost along this step, using the normal stopping power relationship; the coordinates are reset, and the FSE is scattered again, applying the standard Rutherford scattering model of Chap. 3. This procedure is continued until either the FSE leaves the specimen or falls below the minimum energy. If the FSE is found to be below the minimum energy of interest, then we abort the tracking and again estimate the probability that the FSE could reach the entrance surface and escape to be collected as a secondary electron, using the procedure described above.

Figure 8.6A and B shows the program in operation for a 2500-Å foil of silicon at 100 keV. In Fig. 8.6A the fast secondary generation is switched off and the display shows, as usual, the incident electron trajectories. In Fig. 8.6B, FSE generation has been turned on and the display now shows the corresponding FSE trajectories. The distinctive features in the FSE plot are the trajectories of some of the high-energy secondaries that leave almost normal to the original incident beam direction and then travel through the specimen. Only a few secondaries have sufficient energy to travel a great distance, but it is worth noting that the spatial distribution of those that do is quite unlike the usual cone into which the incident electrons spread. An FSE is just as likely to be formed at the entrance surface of the foil as it is at the exit surface, so the FSE produce an approximately cylindrical distribution about the beam axis. This result is of some significance, as discussed below. The low-energy FSE produced do not travel, so their locations mark out the trajectories of the incident electrons. The use of the model is not restricted to thin foils, of course; but

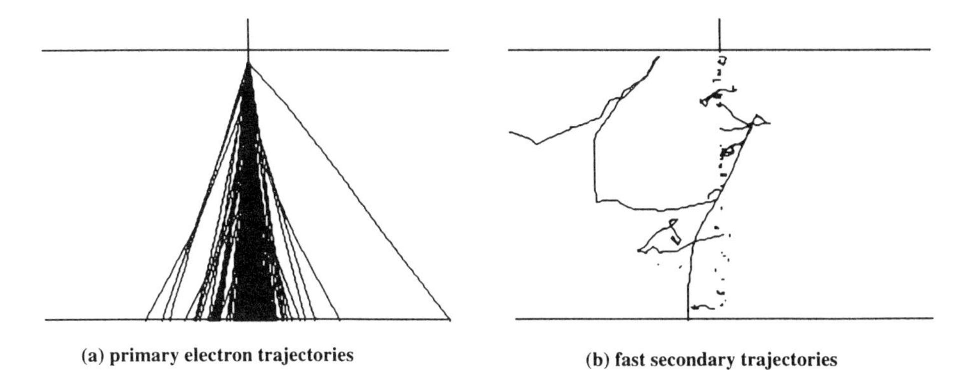

Figure 8.6. Primary (**A** and fast secondary (**B**) electron trajectories in silicon at 100 keV.

because this is two single scattering Monte Carlo models operating together, the computation time for a bulk specimen can be undesirably long. Where possible, therefore, it is desirable to use a thin rather than a bulk sample as the test object. Based on the calculations made in Chap. 6, we can see that an acceptable compromise is to choose a "thin" foil but make its thickness of the order of a third of the electron range. In this case a substantial fraction of the incident electrons are still transmitted, thus saving computation time, but the effective backscattering coefficient is close to that of the bulk material, so the result is a good—though certainly not perfect—estimate of what would be expected for a bulk sample.

The program prints out the yield of secondary electrons that is generated. The magnitude of the value depends on the element, the beam energy, and the assumed thickness of the target, but under all conditions the yield of FSE is quite small, typically in the range 0.001 to 0.1, compared with normal SE yields, which are usually between 0.1 and 1. This is as would be expected, since the high-energy FSE are not "secondary electrons" In the usual definition of the term (i.e., having energies between 0 and 50 eV). This model is therefore only of limited value in predicting the sort of effects that "real" secondary electrons exhibit in the SEM. Nevertheless, some useful studies can be made—for example, in examining the spatial distribution of SE1 electrons at the surface. If we take a sample so thin that no backscattering occurs, then the FSE that are generated are a model of the SE1 electrons. Using the procedure described above, we assume that those FSE that do not reach the surface by the time their energy has dropped below 500 eV subsequently diffuse, following the straight-line approximation [Eq. (8.7)]. The distribution of the secondary electron flux emitted from the surface around the incident beam point can then be determined by dividing the surface into concentric rings of radius r_n. An FSE that is tracked to some radial coordinate r from the beam point and is at a depth z beneath the surface then contributes an SE yield proportional to exp(-z/escape) to the annulus n, where $r_{n-1} < r < r_n$. Figure 8.7 shows the trajectories of the FSE in a 50-Å foil of aluminum at 20 keV and the corresponding SE surface yield profile. The SE1 signal can be seen to be localized with a full width at half maximum intensity of about 3 nm, with the tail of the profile falling away rapidly outside that limit. The width of this distribution does depend somewhat on the material and the beam energy, since at low energies the incident electrons can be elastically scattered close enough to the surface to broaden the distribution, although the lower probability of producing higher-energy FSE will also reduce the intensity of the tail (Joy, 1984). It is the width of this profile that sets an ultimate limit to the resolution of the SEM in secondary imaging mode.

This model also shows (Fig. 8.8) that the SE1 yield rises rapidly with the thickness of the target but then saturates at a thickness that is typically 50 to 60 Å. In the example shown, the SE1 signal reaches about half of its maximum value for a foil thickness of only about 10 to 15 Å, and for very thin films, the SE1 yield is approximately linearly proportional to the thickness. This rapid variation in SE yield

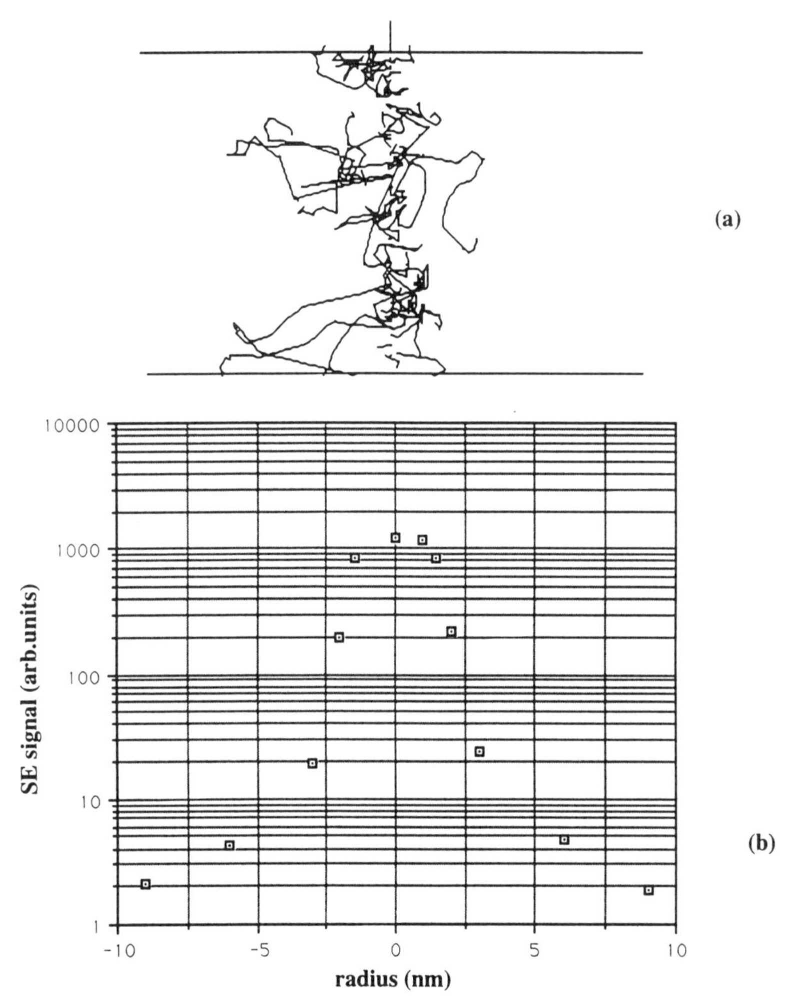

Figure 8.7. Distribution of SE1 electrons in A1 at 20 keV (**a**) and corresponding SE emission profile (**b**).

is exploited in high-resolution SEM (Joy, 1991). A continuous film of chromium or vanadium with an average thickness of about 10 Å is deposited over the sample. In regions where the surface is flat, the SE yield from the film will be that appropriate to the nominal 10-Å thickness, but in areas where there is surface topography, the effective thickness of the film is increased in the beam direction as the film rises up over an edge and thus produces an increase in the SE yield. This "mass-thickness" contrast is an important technique in extending the performance of the SEM.

As can be deduced from an examination of Fig. 8.6, including the FSE contribution makes a very significant change in the way in which energy is deposited in a material. For a thin foil irradiated by high-energy electrons, then, if the FSE are

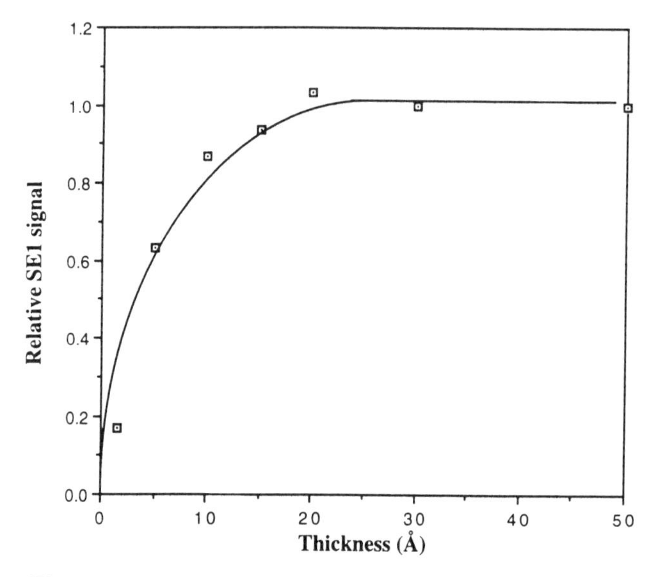

Figure 8.8. Variation of SE1 yield with specimen thickness.

ignored, the energy deposition is concentrated in a tight cone around the beam axis. This would lead to the prediction that effects directly related to energy deposition—such as x-ray generation, the exposure of a resist, or radiation damage—would have high spatial resolution under these conditions. But if the FSE are included, then the situation is seen to be different. The FSE spread the energy deposition perpendicular to the beam direction into a volume that is approximately cylindrical and thus independent of the foil thickness. Because the FSE have, on average, significantly less energy than the primary electrons, their stopping power is much higher and they deposit more energy into the target. Thus, although the yield of FSE is small, the integrated contribution of the FSE to the energy deposition may represent 50% or more of the total (see Joy, 1983, for a detailed discussion). Figure 8.9 shows contours of equal energy deposition computed for the case of an unsupported thin-film resist irradiated at 100 keV. The change in both the absolute magnitude and the distribution of the energy deposition when the FSE are included is very evident. It is this effect which ultimately limits the spatial resoluton of electron beam lithography and x-ray microanalysis at high beam energies, since the average sideways spread of the FSE will become greater than the conventional conical broadening of the incident beam.

8.4 The parametric model

The final approach to computing SE emission is to make a model that correctly describes the experimental behavior of SE without worrying about any of the actual

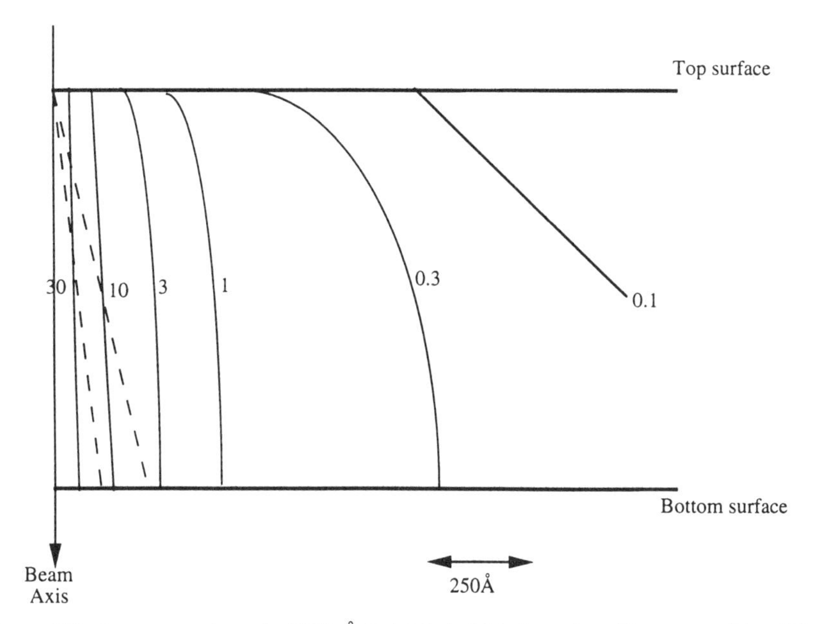

Figure 8.9. Energy contours in 1000-Å PolyMethyl Methacrylate film exposed by 100-keV beam. Dotted lines are corresponding contours for elastic scattering only.

detail of the generation or cascade processes. As before, we have to consider both the generation of the secondaries and their subsequent transport to the surface of the specimen. Salow (1940) and Bethe (1941) independently suggested that the rate N_{SE} at which SE were being produced per unit length of the trajectory as an electron traveled through the specimen was directly proportional to the stopping power of the electron at that point. That is,

$$N_{SE} = -\frac{1}{\epsilon} \cdot \frac{dE}{ds} \tag{8.26}$$

where ϵ is a constant for a given material. This assertion, which is equivalent to saying that a fixed amount of the energy dissipated in a solid is available for SE production, is found experimentally to be a good approximation. Note that if incident electrons are replaced by incident ions, then Eq. (8.26) is still found to be correct—i.e., the SE yields under electron and ion bombardment are in the same ratio as the stopping powers of the electrons and ions in the target (Schou, 1988). Since in any Monte Carlo simulation the stopping power is known for each portion of the trajectory, Eq. (8.26) immediately gives an expression for the instantaneous rate of SE generation.

As before, it will be assumed that the "straight-line approximation" (Dwyer and Matthew, 1985) can be used as a description of the escape of the SE from their point of generation. In this model, an isotropic source of strength N at some depth z beneath the surface would lead to an emission $I(\gamma, z)$ at the surface, where

$$I(\gamma, z) = N.\exp[-z/\lambda\cos(\gamma)] \tag{8.27}$$

where γ is the angle of emission relative to the surface normal. If effects of refraction and reflection on the transmission of SE through the surface are ignored and the SE are assumed to be at zero concentration at the surface, then the average escape probability $p(z)$ for a secondary electron produced at depth z will be (Wittry and Kyser, 1965)

$$p(z) = A.\exp[-z/\lambda] \tag{8.28}$$

where A is a constant of order 0.5 (since half of the SE will move toward the surface and half will move away). The total secondary electron yield δ is then

$$\delta = \int_0^R n(z, E).p(z)\,dz \tag{8.29}$$

where the integral is evaluated from the surface ($z = 0$) to the end of the electron range ($z = R$), assuming that the incident electron travels normal to the surface. Thus combining Eqs. (8.26) and (8.28) gives

$$\delta = \frac{A}{\epsilon}\int_0^R \frac{dE}{dz}\,.\,\exp\left[-\frac{z}{\lambda}\right] dz \tag{8.30}$$

If suitable analytical models are used for the electron range and the stopping power, then Eq. (8.30) can be evaluated to give an expression for the secondary yield in terms of the parameters E, γ, ϵ (e.g., Salow, 1940; Dekker, 1958). Such a result, although obviously an oversimplification, can provide useful insights into both the theory of SE generation (Kanaya and Ono, 1984) and the detail of SE imaging (Catto and Smith, 1973). If, however, Eq. (8.30) is evaluated inside a Monte Carlo simulation where both the instantaneous stopping power and the depth of the incident electron are known at all times, then the model is freed from the limitations of the analytical approximations and becomes both more accurate and more flexible.

Either of the two basic models discussed in this book can be used, although—since SE effects are most usually of interest in bulk specimens—the plural scattering model is usually the most convenient. Incorporating Eq. (8.30) into the model is simple. After the usual computation of the coordinates of the trajectory, end-point coordinates *xn, yn, zn,* then `num_sec`, or the number of SE produced along that step of the trajectory, can be found from Eq. (8.26) in the form

$$\text{num\_sec} = (1/\epsilon).\ [E(k) - E(k+1)]$$

hence the code fragment would be

```
{now calculate the secondary electron signal}
        num_sec:= (E[k]-E[k+1])*se_gen;
```

where `se_gen` is the value of the quantity $(1/\epsilon)$.

The start and finish coordinates of the electron depth beneath the surface along this step are z and zn respectively. If the rate of SE production is assumed to be constant along the step, then, by integration of Eq. (8.28) along the step, the corresponding escape probability $p(z, zn)$ is:

$$p(z, n) = \frac{A\lambda}{(zn - z)} \cdot \left[\exp\left(-\frac{zn}{\lambda}\right) - \exp\left(-\frac{z}{\lambda}\right)\right] \tag{8.31}$$

This integration is necessary because, in general, the step length and hence $(zn - z)$ is much larger than λ. The SE yield, which reaches the surface as a result of the production and escape along this step of the trajectory, is then the product of num _sec and $p(z,zn)$. In the program, this calculation is carried out by the procedure `sei_sig`:

```
Procedure sei_sig;
   {Computes the secondary electron yield along one step of
trajectory}

   var lc, ld:real; {variables local to this procedure}

  begin
          lc:=z/m_f_p;{m_f_p is λ for SE escape}
             if lc>10. then lc:=10.; {error trap}
             if lc<0. then lc:=0.; {error trap}
          ld:=zn/m_f_p;
             if ld>10. then ld:=10;
             if ld<0. then ld:=0;

   {now calculate the signal using generation rate and the integral
 for p(z,zn) - se_yield is the running total of SE production at
 surface}
```

```
    if lc=ld then
  {electron is traveling parallel to surface—don't need to integrate}
  se_yield:=se_yield+0.5* num_sec*exp(-lc)
         else
  se_yield:=se_yield + 0.5*num_sec*(m_f_p/(zn-z))*(exp(-lc)-
exp(-ld));
 end;
```

If the incident electron is backscattered during a step, then the above procedure is not used. Instead, the escape probability is assigned to a random number RND to avoid the necessity of calculating the length of the exit portion of the trajectory that lies within the sample. Since the number of BSE is small, the error produced by this approximation is negligible. Thus the program is modified to read

```
  {test for electron position within the sample}
   if zn<=0 then {electron has been backscattered}
          begin
                 bk_sct:=bk_sct+1.; {count BS electrons}
               {now estimate SE yield for exiting electron}
                 se_yield:=se_yield+num_sec*random;
                     num:=num+1;
                       goto back_scatter;
          end
            else
               sei_sig; {procedure to get increment of SE yield}
               reset_coordinates;
 end;
```

The program SE_MC on the disk implements this code, including, in this case, a loop that changes the incident beam energy through a range of values, so that the variation of SE with energy can be computed.

To run the program, it is necessary to know the values of the parameters ϵ and λ. In general, the most efficient way to obtain these numbers is to measure the total electron yield $(\delta + \eta)$ as a function of incident beam energy from a flat sample and then attempt to fit these data points by supplying the simulation with trial values of ϵ and λ. If a good fit can be obtained, then the validity of the model is established and appropriate values of ϵ and λ have been established. $(\delta + \eta)$ can readily be measured in the SEM by using a calibrated specimen current amplifier and a Faraday cup. If the incident beam current, measured using the amplifier and Faraday cup, is I_b, and if the measured specimen current with the beam on the sample is I_s, then by current balance

$$(\delta + \eta) = \frac{I_b - I_s}{I_b} \tag{8.32}$$

In order to obtain good results, it is necessary to ensure that the specimen does not recollect electrons scattered from the chamber walls or pole piece (e.g., Reimer and Tolkamp, 1980). Figure 8.10 shows measured ($\delta + \eta$) data for carbon, silicon, copper, and silver (Joy, 1993). Since the backscattering coefficient for each of these materials varies only slowly over the energy range, the variation in yield is mostly due to the change in the SE component. In each case the profile has a generally similar form, in which the yield rises to a value of units or greater at around 1 keV. An initial guess for the value of λ can be obtained by using the Salow (1940) result that the SE yield is a maximum when the electron range $R = 2.3\lambda$, where R is evaluated using the normal Bethe expression (see Chap. 4). Estimating λ in this way and initially setting ϵ to 50 eV then allows a trial yield curve to be computed. The quality of fit can then be iteratively improved by adjusting the two parameters. The effects of the choice of λ and ϵ on the yield profile are quite different, so the determination of the best fit values is unambiguous. Table 8.2 gives λ and ϵ values for a variety of commonly encountered elements and compounds.

It is reasonable to regard ϵ and λ as simply adjustable fitting parameters. However, it is also useful to consider the physical significance of these two quantities. ϵ can be considered as being related to the energy required the initiate the SE cascade. This energy will be relatively high, certainly some way above the energy at which the cross section of collisions between an energetic electron and a conduction electron is a maximum—which, for a metal, is the plasmon energy. Thus the values of ϵ deduced from the fitting process, which are mostly 50 eV or higher and so two to three times the typical plasmon energy, are of the right order of magnitude (Joy, 1987a). λ was defined as a characteristic attenuation length which, since an electron cannot be "absorbed," implies that it represents the average distance a SE travels before undergoing a scattering event in which it gives up sufficient energy to make its subsequent escape impossible. λ, which is typically of the order of 1 nm for energies of a few hundred electron volts, falls to a minimum value for energies around 30 to 50 eV and then rises again because there are no large cross-section inelastic scattering processes available (Powell, 1984). Averaging the inelastic MFP over the energy distribution of the secondary cascade predicts values of λ that are quite close to those determined by the fitting procedure. Thus it can be concluded that the values of ϵ and λ derived from a fit to experimental data are physically reasonable and consistent with the cascade model for SE production.

8.4.1 Application of the model

With appropriate values of ϵ and λ substituted in the program SE_MC, it is possible to model some important aspects of the behavior of SE. For example, it is instructive to examine how the secondary yield coefficient δ varies with the angle of beam inclination to the surface, since this is the basis for topographic imaging in the SEM. Figure 8.11 plots the computed SE yield $\delta(\theta)$ at the specimen tilt angle θ, nor-

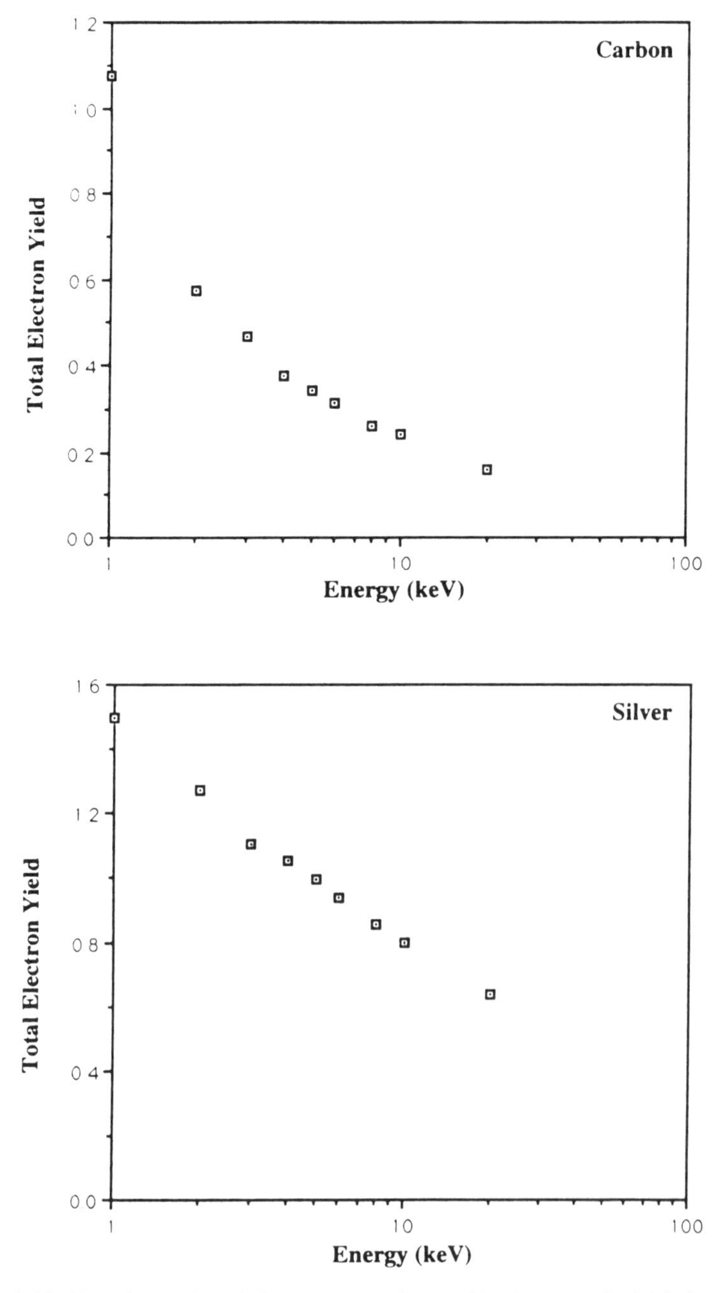

Figure 8.10. Experimental total electron (secondary and backscattered) yields for C, Cu, Ag, and Au.

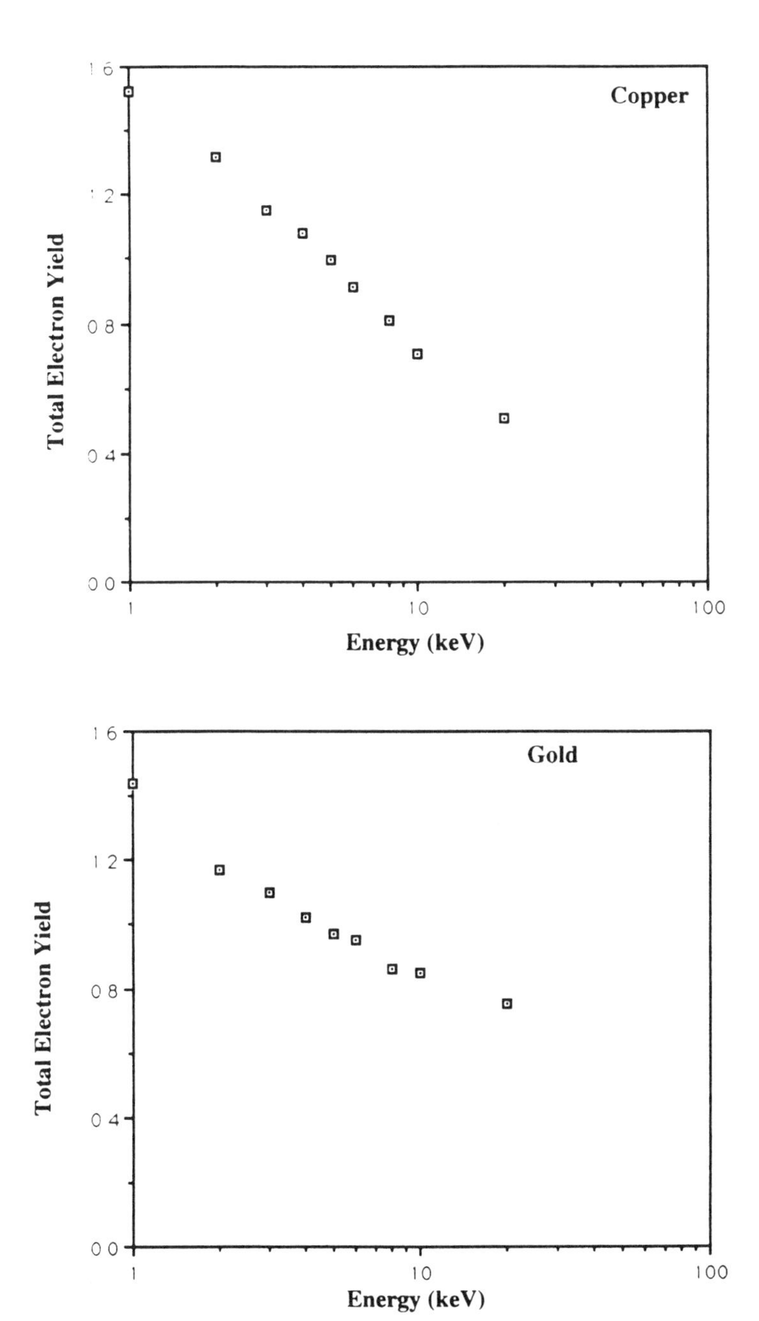

Figure 8.10. (*continued*)

Table 8.2 λ and ε values for a variety of common elements and compounds

Material	ε(eV)	λ(nm)	δm	Em	E2
Carbon	125	5.5	1.8	1.0	3.0
Aluminum	60	2.5	1.67	0.40	1.95
Silicon (xtal)	70	3.0	1.20	0.50	1.45
poly-Si	175	11.0	1.1	0.9	2.5
Chromium	125	2.5	1.16	0.6	2.5
Copper	125	2.5	1.18	0.9	2.8
Molybdenum	100	2.0	1.24	0.65	3.7
Silver	180	3.5	1.0	1.2	4.0
Gold	75	1.0	1.1	0.8	4.6
SiO_2	40	5.0	2.4	0.7	3.5
GaAs	70	5.0	1.2	0.8	3.0
Si_3N_4	110	4.5	1.37	0.8	1.1
PMMA	70	5.0	1.1	0.4	0.6
SiC	60	3.5	1.6	0.6	1.6
Al_2O_3	50	3.0	2.6	0.6	4.0

malized to the yield at normal incidence, for silicon for a number of different beam energies. At accelerating voltages above 5 kV, the yield variation with λ is in close agreement with the measured secant (θ) behavior (Kanter, 1961). As the energy is reduced, however, the variation of yield becomes less pronounced, and at about 0.75 keV a condition is reached where the yield goes through a maximum and then

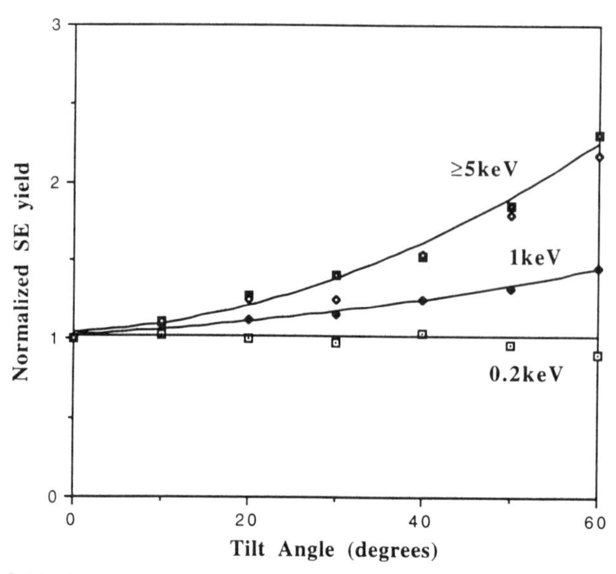

Figure 8.11. Computed variation of SE yield with tilt angle and beam energy.

declines with a further increase in tilt. The decreasing sensitivity of the secondary yield to the beam angle of incidence is readily observed in low-voltage SEM images (e.g., Joy 1989b; Reimer, 1993) and means that the information in a SE image of a surface recorded at low beam energies must be interpreted rather differently to that of the same area recorded at, say, 30 keV. The explanation for this effect is seen to be that at low energies the values of λ, which determines the escape of the SE, is comparable with the incident electron range. Thus the majority of SE generated in the specimen will always escape regardless of the orientation of the surface to the beam.

SE_MC can also be used to examine the charging of a sample under the electron beam. If the incident beam current is I_b, then current balance at the surface requires that

$$I_b = \eta I_b + \delta I_b + I_{sc} \tag{8.33}$$

where I_{sc} is the specimen current flowing to (or from) ground. If the sample is an insulator, then $I_{sc} = 0$ and so charging will occur (i.e., current balance will not be possible) unless $(\eta + \delta) = 1$. In general, there may be two incident beam energies at which this condition may be satisfied, an energy E1, which is typically of the order of 50 to 150 eV, and a higher energy E2, which is usually in the range 500 eV to 2 keV. If the value of E2 for a particular material is known, then operating the SEM at this energy will allow the sample to be examined without charging artifacts even though it is a nonconductor. E2 can readily be found from SE_MC, since both the secondary and backscattered yields are computed. Table 8.1 shows the E2 values computed in this way. These are in good agreement with experimental determinations (Joy, 1987b).

It is also instructive to apply the model to the case of a specimen tilted through some angle θ relative to the incident beam. Both the SE and the BS yield can be expected to rise as θ is raised, so E2 will also be expected to increase. Figure 8.12 plots E2 as a function of θ for a several different materials. In each case E2 rises, but the effect is less pronounced for materials with higher atomic numbers. To a first approximation, we find that E2(θ), the value of E2 computed for angle of incidence of θ, is given by the relation

$$E2(\theta) = \frac{E2(0)}{\cos^2\theta} \tag{8.34}$$

where E2(0) is the value of E2 at normal incidence (Sugiyama et al., 1986; Joy, 1987b; Reimer, 1993). While this result overestimates the variation of E2 for high-atomic-number materials, it is a useful guide as to how E2 varies and is a good example of a result that cannot be produced by a general analytical model of the beam interaction.

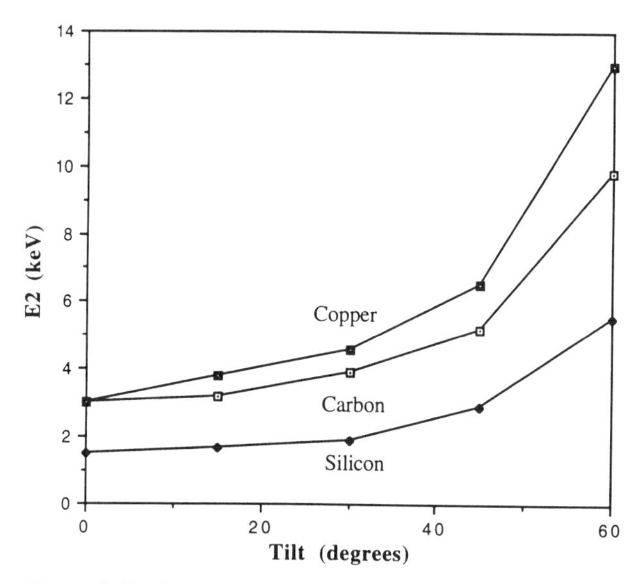

Figure 8.12. Computed variation of E2 with incident angle.

The β factor, which is the ratio of the efficiency of generation of SE by backscattered (BS) as compared with primary electrons [Eqs. (8.16) and (8.17)], can also be determined from this model. The procedure is to compute SE and BS yields as a function of specimen thickness for a given beam energy. As shown in Fig. 8.13, a plot of the SE yield against the corresponding BS yield gives a good straight line,

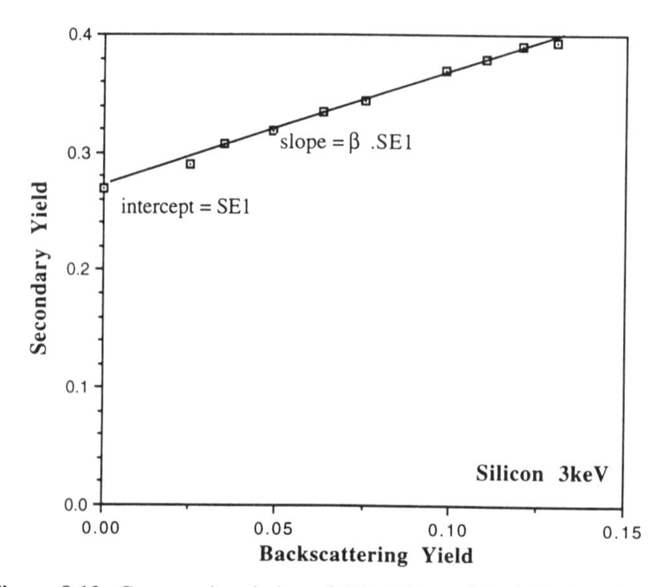

Figure 8.13. Computed variation of SE yield vs. BS yield gives β value.

the extrapolated intercept of which is the SE1 yield and the slope of which is β.SE1. (Note that, as discussed before, if the plural scattering model is being used to carry out this computation, then the sample thickness must be at least 5% of the electron range at the chosen beam energy, so that the simulation includes at least two to three steps as a minimum.) β values are found to be typically in the range 2.5 to 4. This is in good agreement with the values measured by Shimizu (1974) but somewhat lower than the values of 4 to 6 measured by Drescher et al. (1970).

The most important application of this model is as a means to answer the general question: For a specimen of a given chemistry and geometry form, what would be the form of the secondary electron line profile or image that would be produced in a SEM under a specified set of operating conditions? The ability to perform this job is of importance in the field of SEM metrology (Postek and Joy, 1986) as well as being a tool for the detailed interpretation of normal SEM images. The basic requirement is to be able to compute the SE (and usually the BSE) signals at selected points on a surface of arbitrary shape and composition. The first problem is that of defining the geometry of the specimen itself. The mathematical description of a completely general specimen topography is possible through the use of fractal representations, but fortunately this level of complexity is not normally required. In the case of the metrology application, where the specimen is a trench or an interconnect on a substrate, the specimen can be adequately defined as being bounded by a number of straight-line segments that fix the geometry. Computing the BS signal in this case is then relatively straightforward. First, we test whether or not the electron is, at this time, inside or outside of the specimen surface. If it is outside, then its direction of travel relative to the z axis is checked. If the electron is traveling upward (antiparallel to the beam), then it is potentially collectible by a BSE detector and is thus counted as a BSE. If it is traveling downward, then it will eventually impact the specimen again and be rescattered; such electrons are thus not counted.

Determining the SE yield requires a calculation of the escape probability of the secondaries generated within the specimen that takes into account the form of the surface geometry. The first step is to generalize Eq. (8.31) for the escape of a SE. If, as in Fig. 8.14, the electron is traveling in a region bounded by more than one surface—for example, when it is close to an edge or in a raised region such as an interconnect line—the SE escape probability p can be written to a first approximation as

$$p = \sum_{\text{surfaces } i=1}^{i=n} p(s_{i1}, s_{i2}) \tag{8.35}$$

where

$$p(s_1, s_2) = \frac{0.5\lambda}{(s_1 - s_2)} \left\{ \exp\left(-\frac{s_1}{\lambda}\right) - \exp\left(-\frac{s_2}{\lambda}\right) \right\}$$

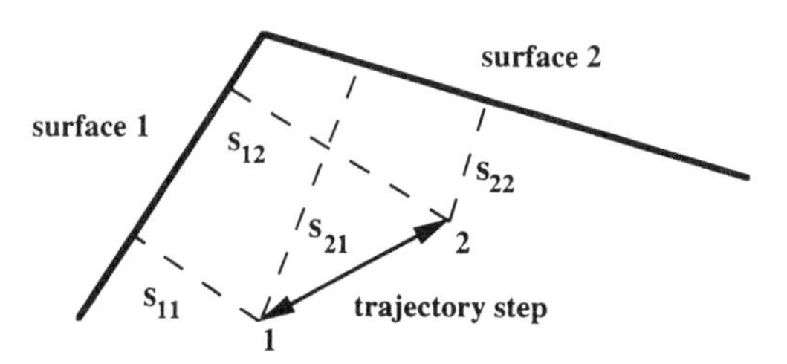

Figure 8.14. Geometry for determination of SE yield.

and the s_{i1} and s_{i2} are, respectively, the perpendicular distances from the start and finish of the trajectory step to the ith surface. It is clear that this approach is an oversimplification because it does not take into account the relative solid angles subtended at the SE generation point by each of the exit surfaces. As a result, the escape probability computed from Eq. (8.35) is an overestimate as compared to values obtained by direct numerical integration (Joy, 1989a) in regions that are close to edges and corners. However, because the exponential terms decay in a distance of order λ, the region that is in error is narrow (i.e., a few nanometers) in comparison with the overall extent of the simulated profile or image (typically a micron or more) and so does not usually present a problem. If higher accuracy is required in these regions, then the values of the directly computed escape probabilities could be substituted.

The program SE_profile on the disk demonstrates the principles of an SE signal line profile simulation, using the procedures discussed in this section. The program is based on the plural scattering code originally developed in Chap. 4. This speeds up the computation, which is desirable, since it is necessary to do a statistically valid number of computations (i.e., 5000 or more) at each of perhaps 20 points to produce a useful profile. On the other hand, the inherent sampling scale of the plural model limits the effective spatial resolution of the simulation to about 2% of the electron range. The conversion to a single scattering model is straightforward and provides a worthwhile improvement in resolution, but at the cost of a greatly increased computation time.

For simplicity, the specimen (Fig. 8.15) is assumed to consist of a parallelipiped structure with a base width of 1 μm placed on a flat substrate, although both the size and shape can readily be changed by obvious modifications to the code. The materials of the structure and the substrate, the height of the structure, and the angles made by the walls of the structure to the vertical can all be determined by the user. The program computes and plots the secondary and backscattered signal profile at 20 different pixel points across the structure and substrate. Because of the symmetrical geometry, all of the pixels are placed in the positive half-space relative to the center of the structure. The computed data are then reflected through the origin of coordinates $Y = 0$ to yield a complete profile. To make the simulation more realistic, the incident electron beam is assumed to be Gaussian in profile and of a

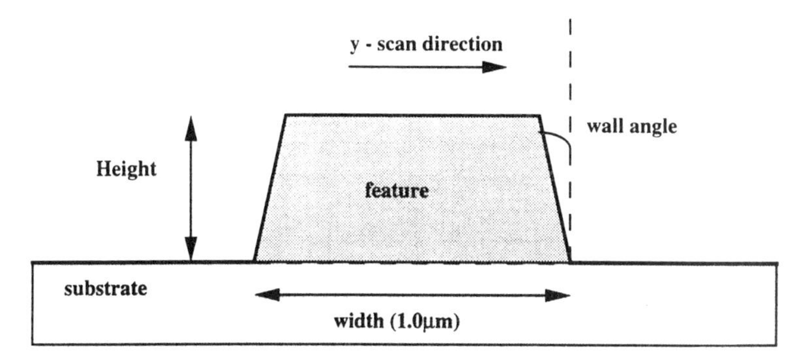

Figure 8.15. Geometry for computation of signal profiles.

chosen finite diameter. The usual assignment at the start of each trajectory, which places

```
x:=0;
 y:=0;
z:=0;
```

is now modified to put

```
x:=x_val;
 y:= y_pos[m] + y_val;
z:=z_val;
```

where the `y_pos[m]` are the nominal values of y as set up in the position array of 20 elements and `x_val  y_val` are Gaussian distributed random numbers with a variance proportional to the desired probe diameter. The routine GasDev performs this computation.

```
Function GasDev:real;
  {generates a Gaussian deviate of unit variance-gliset is a
  global variance indicating if a deviate value is available
  adapted from Press et al., 1986}

var fac,r,v1,v2:real;
 begin      {the function}
      if gliset=0 then   {we need a value}
         begin
           repeat
             v1:=2.0*random-1.0;    {pick a number}
             v2:=2.0*random-1.0;    {pick another number}
               r:=(v1*v1)+(v2*v2); {radius of unit circle}
```

```
          until r<1.0; {points must be in unit circle}
          fac:=sqrt(-2.0*LN(r)/r);    {Box-Muller transform}
        glgset:=v1*fac;               {save one normal deviate}
        GasDev:=v2*fac;               {return the other value}
      end
   else              {gliset=1 so we have a spare value}
      begin
        GasDev:=glgset;               {so return it}
        gliset:=0;                    {reset the flag}
      end;
end;
```

z_val can now be determined from a knowledge of the specimen geometry. A structure defined by its height and width can have two forms (Fig. 8.16)—one of which can be called "overcut" and the other "undercut." If the wall_angle is set to be negative, then the structure is undercut and a Boolean variable (i.e., a variable that can have two values, true or false) is set to false.

```
{compute the parameters necessary to describe the structure}
tn:=sin(wall_angle/57.4)/cos(wall_angle/57.4); {Pascal has no
tangent}
ch:=cos(wall_angle/57.4);    {convert all angles to radians}
  wh:=width/2.0;
  tw:=wh - height*tn;    {from geometry of Fig. 8.16}
   {test for type of structure}
          undercut:=false;    {default condition}
          if tn<0 then undercut:=true;    {test to find actual
          condition}

   {set up some boundary conditions}
       b_edge:=wh;
         if undercut then b_edge:=tw;
```

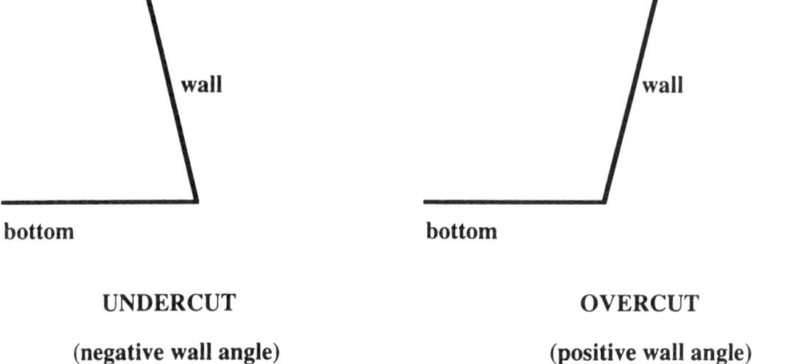

Figure 8.16. Definition of undercut and overcut features of a wall.

```
        plateau:=tw;
          if undercut then plateau:=wh;
```

The y coordinate of the beam is set to one of 20 possible values in an array `y_pos[m]`; so given this value, the geometry of the structure `z_val` can now be determined.

```
   y-val:=y_pos[m] + (probe_size*GasDev); {Gaussian beam profile}

if undercut then   {select appropriate geometry}
     begin
       if y_val>=tw then {it misses the top edge}
        z_val:=0.0
     else
        z_val:=-height;
     end
  else   {it is overcut}
     begin
       if y_val>=wh then z_val:=0
     else
        z_val:=(y_val-wh)/tn: {from geometry of Fig. 8.15}
        if y_val<=tw then z-val:=height;
     end;
```

The rest of the program is straightforward and requires no detailed comment. The longest section is the application of Eq. (8.35) to compute the SE yield. The program proceeds by locating the position of the electron in one of four possible regions—in the substrate not beneath the structure, in the substrate and underneath the structure, in the walls of the structure, in the central region of the structure—and then computing the possible exit paths. One small refinement is to consider the possibility of the specimen recollecting some of its own emitted electrons. Every emerging BSE (e.g., from the side walls of the feature) is checked to see if it has a component of velocity towards the substrate. If it does, then it is allowed to reenter the specimen and the program continues to track it until it is again backscattered or comes to rest.

Figure 8.17 shows SE and BS profiles for the case of an aluminum structure 1 μm wide and 1 μm high with walls at +5° to the vertical (i.e., an overcut structure) placed on a silicon substrate and for beam energies of 2, 10, and 30 keV. The profiles demonstrate how the "image" of this simple object is affected by the choice of beam energy. At 2 keV, both the SE and BSE line profiles reveal the edges of the structure clearly. In the SE image, the simulation shows that the line would appear slightly bright above background, with sharp white edges. The BS profile shows the aluminum line and the silicon substrate at about the same brightness, but the edges of the feature are marked by bright bands with a width of the order of the incident beam range (i.e., about 500 Å). At 10 keV, the SE profile remains much the same, but the

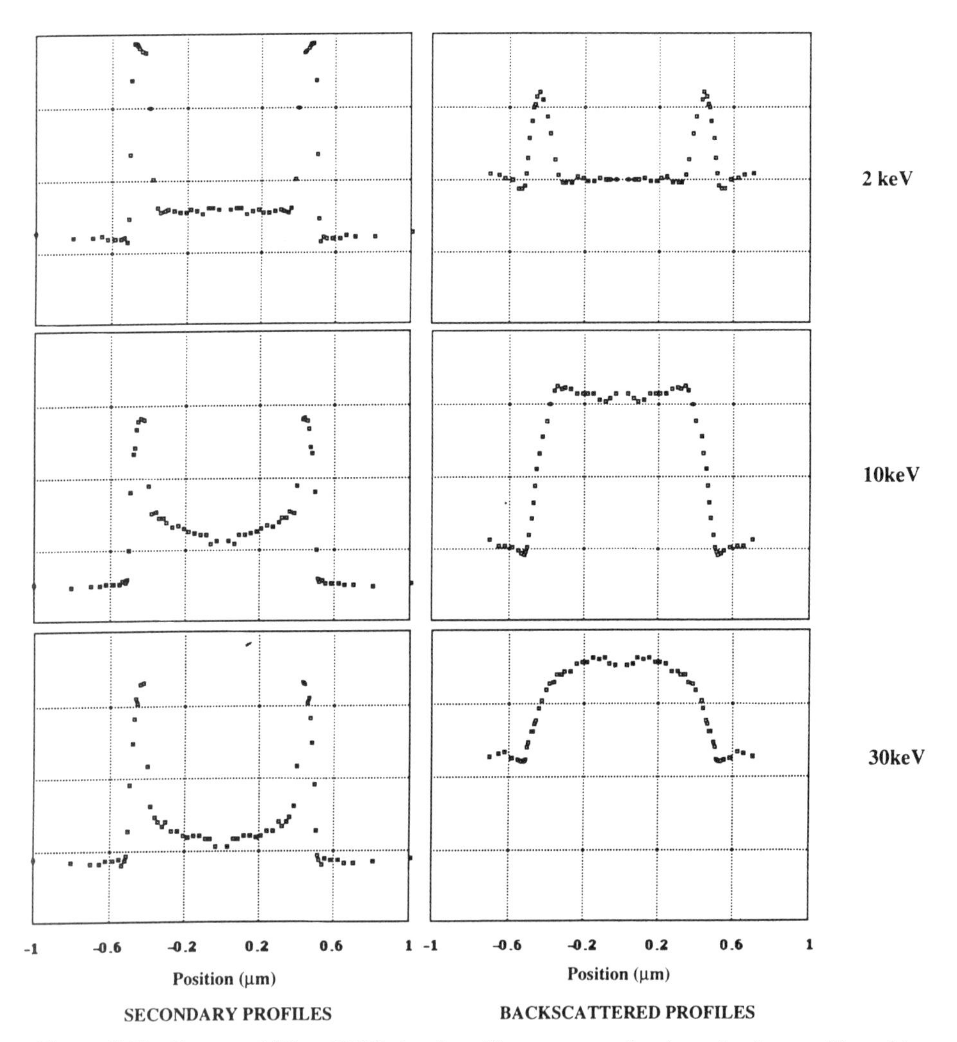

Figure 8.17. Computed SE and BSE signal profiles across an aluminum bar 1 μm wide and 1 μm high, with 5° wall, on a silicon substrate at 2, 10, and 30-keV beam energy.

edge brightness is now greater in magnitude. The BS profile has changed substantially, however, because the interaction volume of the beam is of the same order of magnitude as the size of the feature. The beam range is now of the order of the width of the feature, so the whole line is bright above background with no clearly defined edges. Note also that the background intensity from the silicon substrate dips as it approaches the edge of the feature because of electron penetration beneath the structure. Finally, at 30 keV, the SE and BSE images are similar to those at 10 keV, with the SE profile dominated by the edge bright lines and a low contrast in the BSE profile because the thickness of the line is now a small fraction of the electron range.

A comparison of the various line profiles with the real cross-sectional geometry of the structure shows that determining the position of the top and bottom edges from any of the traces is difficult or impossible. Hence, although it is easy to image a micron-scale device, "measuring" the profile with any accuracy using the SE or BSE signals is a much more challenging task because of the constantly shifting relationship between the real and apparent edge positions. One of the most useful applications of such simulations as these, therefore, is to try and devise rules that will allow experimental line profiles to be interpreted in such a way that the width, height, and wall angle of the structure from which they came can be deduced (see, e.g., Postek and Joy, 1986).

A comparison of these computed profiles and corresponding experimental data shows that the qualitative agreement is excellent, with the simulations predicting all of the features of the experimental profiles. However, the level of quantitative agreement is not always so encouraging, for a variety of reasons. A key reason is the behavior and efficiencies of the SE and BSE detectors used in the experimental system. In the computation, it is assumed all secondaries are collected. In practice, the detection efficiency of the SE detector will depend on where it is relative to the irradiated area, on the bias applied to it, and on the presence and magnitude of any local charging. These considerations can drastically modify the detail of the SE profile. Unfortunately, computing the efficiency of the SE detector requires a determination of the electrostatic field distributions in the SEM specimen chamber, and this is a lengthy task (Suga et al., 1990; Bradley and Joy, 1991, Czyzewski and Joy, 1992). In addition, in complex structures such as parallel arrays of closely spaced lines, there is a possibility that emerging BSE will be recollected by some other part of the structure, generating SE at the point of entry and during the subsequent travel (Kotera et al., 1990). Detailed calculations, therefore, require testing for the possibility of recollection by tracing the path of the BSE as they leave.

Expanding these line-profile computations to the simulation of a complete image is, in principle, straightforward, although the time required to obtained adequate statistics might be daunting on any but the fastest computers. Nevertheless, with the growing interest in quantitative image interpretation (e.g., for microdimensional metrology), and for the comparison of images between, say, the SEM and the scanning tunneling microscope (STM), this is likely to be an important application and extension of the techniques discussed above.

9

X-RAY PRODUCTION AND MICROANALYSIS

9.1 Introduction

The process of fluorescent x-ray generation in a sample by electron beam irradiation makes it possible to perform a chemical microanalysis of small (micrometer size or less) regions of a specimen. However, in order to obtain quantitative data, it is necessary to be able to describe in considerable detail the distribution—both laterally and in depth—of the x-ray production. It was this need that led to the initial studies of Monte Carlo modeling by Bishop (1966). Although, it many cases, simple analytical models have subsequently been developed that allow many of the required results to be obtained with adequate accuracy, only Monte Carlo methods offer a general approach to the microanalysis of an arbitrary specimen. In this chapter we examine x-ray production, both characteristic and continuum (*Bremsstrahlung*), in thin foils and in bulk specimens using the Monte Carlo models previously developed.

9.2 The generation of characteristic x-rays

The production of characteristic x-rays along a step of an electron trajectory can be calculated if the cross section for x-ray production is known. Typically the Bethe cross section for inner-shell ionization is used, in the form:

$$\sigma = 6.51 \times 10^{-20} \cdot \frac{n_s b_s}{U\, E_c^2} \log(c_s U) \tag{9.1}$$

where n_s is the number of electrons in the shell or subshell, E_c is the critical ionization energy in kilo electron volts, U is the overvoltage E/E_c, b_s and c_s are constants, and σ then has the units of ionizations per incident electrons per atom/cm^2. For example, for the K shell, b_s and c_s are 0.9 and 0.65 respectively (Powell, 1976) for the overvoltage range $4 \leq U \leq 25$. For the L and M shells, the appropriate constants and range of applicability are less well established (Powell, 1976).

In some cases, for example, the production of x-rays in a thin foil, the energy of the electrons may always be high enough to put U in the correct range; but in the case of bulk specimens, it is clear that, since the energy of the electron decreases as it travels through the specimen, ultimately the overvoltage will no longer satisfy the condition $U \geq 4$ and the predicted cross section may become inaccurate. Alternative cross-section models are available, for example Casnati et al. (1982), which are valid for low overvoltages, and these should be considered for the most accurate work; but the familiarity and convenience of the Bethe expression has led to its nearly universal use in all types of conditions, whether warranted or not.

The number of x-rays I_s per incident electron produced along a step of the trajectory is then

$$I_s = \sigma N_A \, \rho \, \omega. \, \text{step}/A \tag{9.2}$$

where N_A is Avogadro's number (6×10^{23} atoms/mole), A is the atomic weight, ρ is the density, ω is the fluorescent yield, and the step length is in centimeters. Since A, N_A, ρ and φ are constant, it is convenient to extract these quantities, and the constants ahead of the functional variables in Eq. (9.1), and evaluate I_s in the form

$$I_s = \text{step} \cdot \frac{\log(c_s U)}{U \, E_c^2} \tag{9.3}$$

only restoring the numerical constants when an absolute yield value is required.

9.3 The generation of continuum x-rays

In addition to the characteristic x-ray signal used for microanalysis, there is also a continuum, or *Bremsstrahlung*, x-ray signal generated by the slowing down of the electrons in the coulomb field of an atom. By analogy with Eq. (9.2), we can write the continuum yield per incident electron into a unit steradian I_{co} as

$$I_{co} = Q \, N_A \, \rho. \, \text{Step}/A \tag{9.3}$$

where Q is the continuum cross section. The continuum radiation, unlike the characteristic line, is anisotropicaly emitted, being peaked about the forward direction of travel of the electron. In considering *Bremsstrahlung* production in bulk samples, this effect is of little consequence because it is averaged out by the plural scattering of the electrons; but in the case of thin films, where the incident electrons are only scattered through small angles, the polarization and directivity of the continuum must be properly accounted for.

A simple cross section for continuum production can be obtained by combining experimental observation and some theoretical results (Fiori et al., 1982). The maximum energy that can be given up by an incident electron is equal to its kinetic energy E_0, numerically equal to the beam voltage V_0; thus the highest energy that can appear in the continuum spectrum is E_0, the so-called high-energy or Duane-Hunt limit. Second, for thin specimens it is found that the continuum energy in an energy interval ΔE is about constant (Compton and Allison, 1935) from zero up to the high-energy limit. Thus, the fraction of the total emitted continuum energy E_T in the energy range E to $E + \Delta E$ is $\Delta E/E_0$. The efficiency of the generation of continuum production is defined as the total continuum energy in the range from near zero to E_0, generated by electrons that lose an amount dV of their energy divided by the energy lost. Kirkpatrick and Wiedmann (1945) showed from the theory of Sommerfeld (1931) that this efficiency was $2.8 \times 10^{-9}\, Z\, V_0$, where V_0 is the beam voltage in kilovolts and Z is the atomic number. The total continuum energy is thus

$$E_T = 2.8 \times 10^{-9} Z\, V_0 . dV \tag{9.4}$$

The fraction of this quantity in the energy interval ΔE then gives the number of photons I_{co} of energy E; i.e., the number of photons in the energy interval ΔE is $E_T . \Delta E/(E\, E_0)$. So

$$I_{co} = 2.8 \times 10^{-9} (Z/E) \Delta E . dV \tag{9.5}$$

In the Monte Carlo simulation, dV is simply the energy loss occurring along a given step of the trajectory, a quantity that is evaluated either directly from the stopping power (in the single scattering model) or as the difference $E[k] - E[k + 1]$ in the plural scattering model, so Eq. (9.5) can be used to give the continuum intensity at some energy E in the energy interval ΔE directly.

This simple expression produces quite useful results in many cases of interest, but it must be realized that it is a drastic simplification, since it assumes both isotropic emission and uniformity in energy distribution. More accurate and detailed models are available, the most widely used of which is that of Sommerfeld (1931) as modified by Kirkpatrick and Wiedmann (1945). Sommerfeld's model assumes a pure Coulomb field about a point nucleus ignoring screening effects and represents the scattered electrons as plane waves. Stricly speaking, this theory is valid only for low electron energies, where relativistic effects can be neglected. Kirkpatrick and Wiedmann (1945) gave an algebraic fit to the Sommerfeld theory that correctly accounts for the polarization of the *Bremsstrahlung* and is more easily computed numerically. The algebra is messy and will not be reproduced here, but a routine for the evaluation of the Kirkpatrick and Wiedmann expression based on that given by Statham (1976) will be given later in this chapter.

9.4 X-ray production in thin films

9.4.1 Spatial resolution

The simplest case of interest is that of determining the spatial resolution of x-ray microanalysis. This problem is of importance in analytical electron microscopy (AEM), where x-ray analysis is used to determine the composition of small precipitates or the composition profile at a boundary. In this kind of situation, the specimen is thin (typically 300 to 1500 Å) and the incident beam energy is high (100 to 200 keV). As can be seen by running the SS_MC program (Chap. 3) for this type of condition, the majority of electrons pass through the sample unscattered because the specimen thickness is comparable to and smaller than the elastic mean free path (MFP) at this energy. Since, in addition, the beam energy is usually much greater than the excitation energy of the x-ray line of interest, the volume associated with x-ray generation is identical with that associated with the electron scattering, since the ionization cross section will be essentially constant. Since the x-ray spatial resolution is commonly defined as the radius of the volume within which 90% of the emitted x-rays are generated, in this approximation the resolution can be taken as being the radius within which 90% of the transmitted electrons emerge from the bottom surface of the foil. Incorporation of this into the SS_MC program is straightforward. If an electron is determined as being transmitted (i.e., $zn >$ thickness of the foil) then the length of the exit vector is first found and the exit radius from the beam axis ($x = y = 0$) is computed. The bottom surface is divided into an array of 100 annular rings, `radius [0 . . 99]`, of constant width `scale` where `scale` is set to 10 Å. The exit radius is converted to give the number of the annular ring (0 to 100) through which the electron left, and one count is added to the total for that ring.

```
    {after determining that this electron is leaving}
   {find length of vector from x,y,z to bottom surface}
              ll:=(thick-z)/cc
  {hence the exit coordinates on the bottom surface are}
              xn:=x+ll*ca;
              yn:=y+ll*cb;
       {and the exit radius about the beam axis is}
              radial:=sqrt((xn*xn)+(yn*yn));
       {convert this to an integer identifying the annular ring}
              r_val:=trunc(radial/scale);
        {and use r_val to index the array and add one count}
         if r_val<100 then {will not overflow the array
              bounds}
              radius[r_val]:=radius[r_val] + 1;
                 goto exit;
         end;
  {otherwise perform another trajectory step etc. . . . . . .}
```

At the end of the desired number of trajectories, the beam is found by starting from the center annulus (r_val=0) and counting out until 90% of the total number of transmitted electrons have been included. The radius at which this occurs is then the beam broadening figure.

```
            {now calculate the 90% beam broadening radius}

                        dum:=0    {dummy summing variable}

   {total number of transmitted electrons is traj_num - bk_sct. We
     have to reach 90% of this number}
              broad:=trunc(0.9*(traj_num-bk_sct));
     {compute the running sum as the radius is increased}
               for k:=0 to 99 do
                 begin
                   dum:=dum + radius[k];
          if dum>=broad then {we have reached 90% at this value of k}
                     goto test; {so stop counting}
                 end;
          test:                  {exit from summing loop}
   {and display the computed beam broadening as a line of appropriate
length}
          MoveTo(center-trunc(k*scale*plot_scale), bottom+15);
          LineTo(center-trunc(k*scale*plot_scale), bottom+15);
                   readln; {freeze the display}
```

Figure 9.1 shows how the beam broadening computed in this way compares with the simple analytical estimate given by Goldstein et al. (1977) for the case of a 100-keV beam incident on a foil of copper. It can be seen that the two models predict the same type of behavior with increasing thickness, even though they give somewhat different values for the broadening. This is not surprising, because the analytical model is estimating the average exit diameter on the assumption that each electron is scattered just once, at the midpoint of the foil, and this figure is not directly related to the 90% definition used in the Monte Carlo case. In addition, once the foil becomes thick enough for plural scattering to be significant, the value estimated by the analytical model is no longer valid.

It is clear, in any case, that a single "beam broadening" number cannot faithfully represent the actual experimental situation. While it provides a guide as to the resolution to be expected, it cannot say what effect this resoluton might have on, for example, a measured composition profile. For that kind of problem, it is necessary to use the actual x-ray generation profile of the beam, again on the assumption that this is directly related to the distribution of trajectories of the incident electrons. The program AEM_MC on the disk provides this ability. It is essentially the same single scattering program, the only modification being that the incident electron beam is

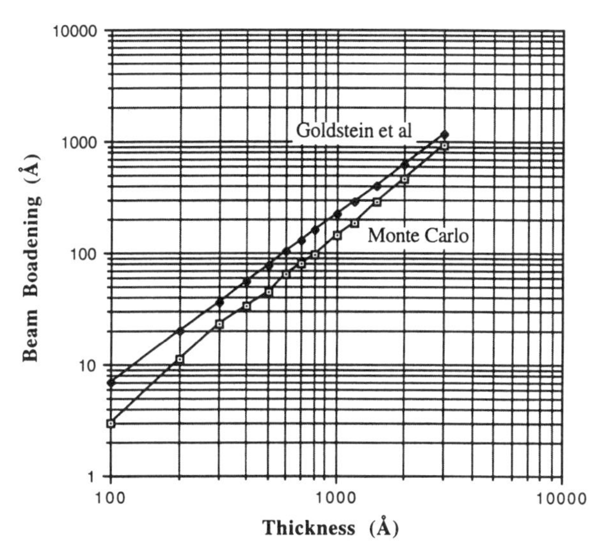

Figure 9.1. Beam broadening in copper at 100 keV using Monte Carlo method and Goldstein et al. (1977) equation.

assumed to be a Gaussian distribution of some given full width at half maximum height (FWHM). Basically, the first part of the procedure is identical to that given above, so that the number of electrons traveling through—and hence the number of x-rays produced in a given annular ring—is determined as before. Imagine now that the beam is placed at a distance X from some test point in the sample. When X is very large and negative, no x-rays will be produced at the test point, but the number will increase as X is reduced, reach a maximum when $X = 0$, and then fall away again as X becomes large and positive. A plot of the integrated x-ray generation in the positive half space $x > 0$ as a function of the incident beam position X gives a S-shaped profile (fig. 9.2), which rises from 0% to 100% of the total x-ray generation. In a manner analogous to that used for specifying the probe diameter of an electron beam, the distance travelled in raising the x-ray signal from 10% to 90% of its maximum value can be called the x-ray spatial resolution. The program plots the profile and marks on the 10% and 90% levels so that this number can be measured. Although this value is again a single number, the visual profile provides a guide as to what can be expected to happen. When the specimen is thin, the profile is dominated by the Gaussian shape of the probe, since electron scattering is small and the profile is a standard error function (or "erf") curve. As the foil thickness increases, however, the profile changes shape and is composed of two distinct regions, (1) a central region that is still approximately the error function, surrounded by (2) broad tails as a result of electron scattering.

The effect of this particular beam profile on an experimental measurement can

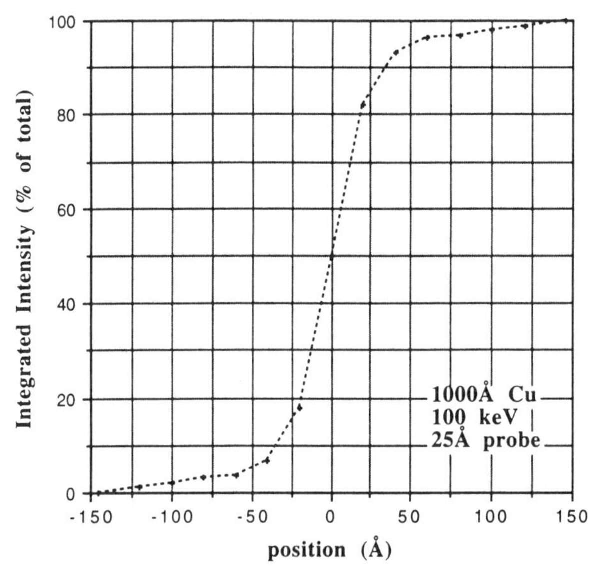

Figure 9.2. Variation of integrated x-ray intensity with position.

now be observed. The program permits two kinds of situations to be considered (Fig. 9.3): the case in which two materials A and B are in contact at some interface and in which B is diffusing into A, and the case of an interface region of some width W of material B surrounded by material A, again allowing for the diffusion of B into A. It is assumed in either case that the electron scattering powers of A and B are sufficiently similar that the form of the electron beam profile is not significantly altered. $C(B)_x$ the concentration of B in A at some distance x from the interface, is given the form:

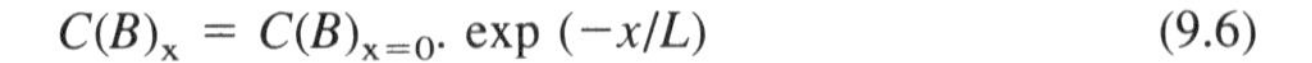

$$C(B)_x = C(B)_{x=0} \cdot \exp(-x/L) \tag{9.6}$$

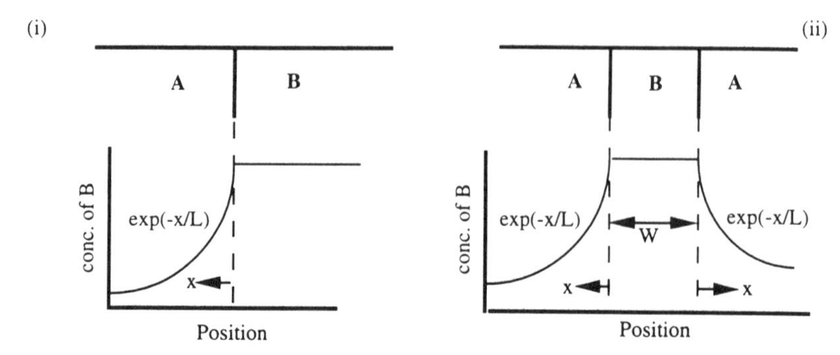

Figure 9.3. Interface geometries for thin-film x-ray program.

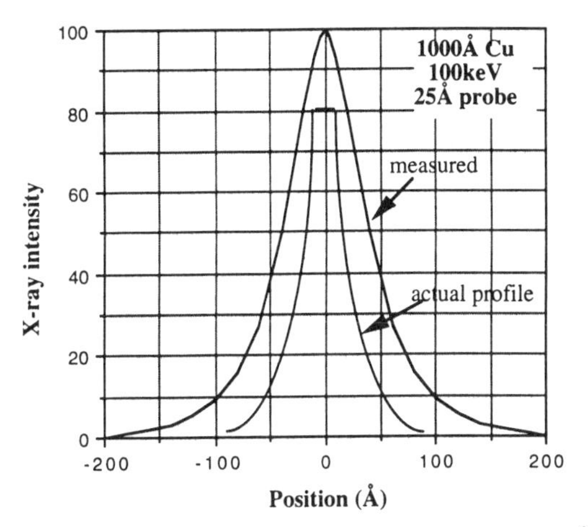

Figure 9.4. Comparison of computed and actual profiles across a 10-Å-wide boundary in copper, with $L = 30$ Å.

where L is a characteristic diffusion length. If L is specified, then the form of the composition profile that would be experimentally measured across the chosen interface can be found by numerical convolution of the computed beam profile with the given form of the interface. Because the procedures for doing this are straightforward, they are not given here, but they can be studied in the source code on the disk. Figure 9.4 shows how the "real" composition profile and the predicted measured profile would compare for a 1000-Å-thick copper film and a 25-Å beam at 100 keV for the case where the interface region was 10 Å wide (e.g., a grain boundary) and the diffusion length L is 30 Å. If experimental data are available from an analytical electron microscope, the values of L and W can be obtained by iteratively fitting the computed and measured profiles. The model can readily be extended to other geometrical situations.

9.4.2 The effect of fast secondary electrons

In the previous chapter, the generation of fast secondary electrons (FSE) was incorporated into a Monte Carlo program. It was noted that these FSE, which have energies up to half that of the incident beam, tended to travel almost normal to the original direction of the incident beam. Clearly, if these FSE can generate x-rays, this could adversely affect the spatial resolution of microanalysis. Equation (9.3) shows that the yield of x-rays by an electron of energy E is proportional to $1/U$, where U is the overvoltage E/E_k. Consider then the case of oxygen x-rays ($E_k = 530$ eV) being fluoresced by incident 100-keV electrons, and by, say, 2-keV fast secondary electrons. The efficiency of oxygen x-ray ionization by the FSE is 50 times higher than the efficiency for the primary electron, so even if only one primary electron in a hundred generated an FSE, 50% of the measured x-ray yield could be coming from secondaries rather than the incident beam. Since, as we saw, the spatial distribution of the FSE is also quite different from that of the incident beam, it is clear that the effect of including fast secondary production could be significant.

To examine this effect, the Monte Carlo model must be modified to include x-ray generation using Eq. (9.3) for some arbitrary x-ray energy. The code for this is straightforward:

```
Procedure generate_x-rays (energy:real;stepsize:real);
   {computes the x-ray yield using Eq. (9.3) given the electron
energy and the critical excitation energy of the x-ray. Fixed
constants can be inserted later if required}

var
   x_ray_yield:real;
    at_z, at_r:integer;
  begin
            x_ray_yield:=ln(energy/E_crit)/(energy*E_crit);
            x_ray_yield:=x_ray_yield*stepsize;
   {assign X-ray production to a box at given radius and depth}
              at_z:=round((z+zn)/(2*step)); {depth}
   at_r:=round(sqrt((x+xn)*(x+xn)-(y+yn)*(y+yn))/(2*step)); {radius}

              if at_z<=0 then at_z:=0;    {to protect array}
   if ((at_z<=50) and (at_r<=50)) then {within bounds so put into
array}
   x_ray_gen[at_z,at_r]:=x_ray_gen[at_z,at_r]+x_ray_yield;
  end;
```

In this example, the computed x-ray yield is deemed to have occurred at a point defined by the midpoints of the trajectory step $(x + xn)/2$,$(y + yn)/2$, and at a depth $(z + zn)/2$. The data are stored in an array `x_ray_gen[0 . . 50,0 . . 50]` formed of annuli whose width and depth are defined by the quantity `step`. The box size can be chosen to be any value, but a convenient choice is often to make `step` equal to one-fiftieth of the estimated electron range. The variable `stepsize` passed to the procedure is the length of the trajectory step. If, on this step, the electron is either transmitted through or backscattered from the specimen, the step-size must be modified accordingly. This can be done exactly—e.g., the actual distance `ll` traveled within the specimen for an electron that is backscattered out of the specimen from some point z beneath the surface of the specimen is $ll = z/cc$, where cc is the usual direction cosine. Otherwise the stepsize can be approximated putting the stepsize equal to stepsize*RND.

The insertion of the procedure into the FSE_MC code is obvious and needs no detailed comment. In either the primary loop, or in the FSE loop, the x-ray genera-

Table 9.1 Effect of FSE on X-ray resolution

Line Used	Mode	Resolution	X-ray Yield
Iron Kα	No FSE	60 Å	—
	With FSE	64 Å	+1%
Silicon Kα	No FSE	60 Å	—
	With FSE	65 Å	+3.1%
Oxygen Kα	No FSE	60 Å	—
	With FSE	75 Å	+6.0%
Carbon Kα	No FSE	60 Å	—
	With FSE	85 Å	+11.2%

tion can be computed as soon as the new coordinates *xn,yn,zn* have been calculated and the subsequent behavior of the electron has been determined (i.e., does the electron remain in the sample or is it transmitted or backscattered?). Thus

```
....................
zn:=z+ step*cc;
{get x-ray production now for electron remaining in sample}
if E>E_crit then {ionization can occur so}
generate_x-rays(E,step);
```

The effect of including FSE production is often quite significant. Table 9.1 shows data computed for the effective resolution and total x-ray yield for several different elements in a 500-Å thick sample of iron examined with a 100-keV beam.

It can be seen that in the simple approximation where secondary effects are neglected, the spatial resolution is independent of the energy of the x-ray line that is being examined, because only the volume occupied by the trajectories of the incident electrons are considered. But when FSE are included, the resolution gets worse because of the lateral motion of the secondaries. The lower the energy of the x-ray, the further a given secondary can travel before it no longer has sufficient energy to excite the chosen line, so as the ionization threshold is lowered, the computed resolution becomes worse. For iron Kα where E_{crit} is around 7 keV, the change in resolution is negligible because relatively few FSE have high enough energies; but for carbon Kα (E_{crit} = 0.28 keV), the resolution is degraded by some 40%. Similarly, including FSE production also changes the predicted yield of x-rays from an element. The extent to which this occurs depends on the nature of the matrix in which the element to be observed is sitting. If the matrix is of low density and low atomic number, then the number of FSE produced, and the consequent change in x-ray yield, will be small; but for a dense, high-atomic-number matrix, the number of FSE produced is high and consequently extra fluorescence of the element occurs. This indicates that care must be used in performing quantitative microanalysis under conditions where FSE production is significant. For example, at high beam energies and thus generally high overvoltage-ratios, the majority of x-ray production may be coming via the FSE rather than from the incident electron. In this case the spatial resolution will be significantly worse than predicted by the simple theory—and, in fact, will get worse rather than better as the beam energy is increased, but it will be more or less independent of the foil thickness—and the intensity ratios of various elements will vary in a complex way with composition because changes in chemistry will cause changes in the number of FSE and hence in the amount of excitation. Microanalysis at beam energies of 200 keV or above is likely to demonstrate these problems.

9.4.3 Peak-to-background ratios

A common measure of the quality of performance of an AEM is the peak-to-background ratio that can be achieved for a given x-ray line, typically chromium (Williams and Steel, 1987). Values that have been published in the literature for various microscopes vary by nearly a factor of 10, and there has also been considerable discussion about how the figure might be expected to change with the accelerating energy of the microscope. The Monte Carlo simulation can be modified to address some of these questions. The simplest approach is to use a Kramer's law model for the *Bremsstrahlung* contribution in the form given in Eqs. (9.4) and (9.5). Although this is only an approximation, in particular because it ignores the anisotropy of the continuum, it provides a good starting point for a more detailed analysis. In order to make use the routine for both electron transparent and bulk samples, it is convenient to calculate the depth distribution of the continuum signal in the chosen energy window, since this will be needed later on. By analogy with the $\phi(\rho z)$ ("phi_ro-z") depth distribution of characteristic x-ray production, this can be called bkg_ro_z. The code would have the following form:

```
Procedure continuum (energy:real;stepsize:real);
   {this computes the Bremsstrahlung flux per energy step using a
   Kramer's law model. No account is taken of the anisotropy of radi-
   ation. The continuum window is at an energy E_bkg}
constant
  window=0.01; {width of continuum energy window in keV}
var
     bkg_yield:real;
     del_E:real;
          begin
      position:=round(50*z/thick); {divide thickness into 50 steps}
               {now compute the continuum signal}
  if energy>E_bkg then {the continuum window can be excited}
                     begin
  del_E:=stepsize*stop_pwr(energy)*density*1E-8; {energy lost along
  step}
```

```
  bkg_yield:=2.8E-9*(at_num/E_bkg)*del_E*window; {Kramer's law}
                        end
                 else
bkg_yield:=0;
if ((position>0) and (position<=50)) then {we are within bounds of
array}
          bkg_ro_z[position]:=bkg_ro_z[position]+bkg_yield;
     end;
```

The routine computes the Kramer's law contribution, assuming that the energy window used for measuring the *Bremsstrahlung* is 10 eV (0.01 keV) wide. The rate of continuum production is proportional to the electron stopping power, so the energy lost along this step of the trajectory must be determined. Note that this is the flux emitted into 4π steradians. The angular distribution of the continuum is not, however, isotropic, but forward peaked about the beam direction through the thin film, so some account must be taken of this. The corresponding characteristic intensity can be computed by a slight modification of the routine given above, so as to give the depth distribution (phi_ro_z).

```
Procedure generate_x-rays(energy:real;stepsize:real);
   {computes the x-ray yield using Eq. (9.3) given the electron ener-
   gy and the critical excitation energy of the x-ray. Fixed con-
   stants can be inserted later if required}
   var
      x_ray_yield:real
       position:integer;
  begin
          if energy>E_crit then {x-rays will be produced}
          begin
             x_ray_yield:=ln(energy/E_crit)/(energy*E_crit);
             x_ray_yield:=x_ray_yield*stepsize;
           end
              else
                x_ray_yield:=0;
      {assign x-ray production to a given depth}
              position:=round(50*z/thick); {depth counter}
      if ((positon>=0) and (position<=50)) then {within bounds of
      array}
phi_ro_z[position]:=x_ray_gen[position]+x_ray_yield;
  end;
```

Again, this is the intensity produced into 4π steradians. Since we want the peak-to-background intensity ratio at a given energy, E_Bkg for the continuum can be put equal to E_Crit for the characteristic line. Then adding the lines:

```
   {this computes the characteristic and continuum signals}
              generate_x-rays(s_en,step);
              continuum(s_en,step);
```

to the program at appropriate points in the program (i.e., when the coordinates are reset, or when the electron is backscattered or transmitted) calls the two routines. At the end of the simulation, the depth distributions for the characteristic and continuum signals must be summed up over the whole thickness of the specimen. Since both signals are at the same energy, it is a reasonable approximation to assume that any x-ray absorption in the specimen will be the same in both cases and thus will disappear in the ratio. Similarly, the solid angle subtended by the detector is the same for both the continuum and characteristic signals and also disappears from the result. Finally, the missing constants (such as Avogadro's number, density, and atomic weight) that appear in the original equations but not in the code must be reinserted so as to give the proper numerical values.

```
{sum the x-ray contributions}
      for k:=0 to 50 do
        begin
          the_peak_is:=the_peak_is + phi_ro_z[k];
          the_bkg_is:=the_bkg_is + bkg_ro_z[k];
        end;
{put back the missing constants}
the_peak_is:=the_peak_is*6.5E-20*1.2*6E23*(density/at_wht)*1E-8;
        the _bkg_is:=the_bkg_is *1E4;
    {so the computed peak-to-background ratio is}
        p2b_ratio:=the_peak_is/the_bkg_is;
```

The program "P_to_B" on the disk implements all of this code. Choosing chromium as a specimen, the predictions of the program can be tested against published data values (Williams and Steel, 1987) for the "Fiori" number—the ratio of the integrated intensity in the Cr Kα line to the intensity in a 10-eV-wide window of the continuum at the same energy. At 100 keV, the program predicts a value between 1300 and 1200, falling slightly as the film thickness is increased from 250 Å to 5000 Å. Figure 9.5 plots how this peak-to-background (P/B) ratio varies with incident beam energy for a 1000-Å-thick film of chromium. At 10 keV, the ratio is only about 250, but this value rises rapidly as the energy is increased (and plural scattering becomes less pronounced in the sample). Note that the P/B ratio continues to rise for energies above 100 keV, but only relatively slowly. These values are in fairly good agreement with, although somewhat lower than, both published experimental data and other theoretical estimates (e.g., Fiori et al., 1982). This level of agreement is only as good as might be expected, since the model is not accounting for the forward-directed anisotropy of the continuum and so is probably overestimating its intensity in the direction of the detector. The simple Bethe cross section for the characteristic signal is also of uncertain accuracy at 100 keV and above.

To obtain a more accurate result, it is necessary to use a continuum cross section, which is not only more accurate but which explicitly accounts for the

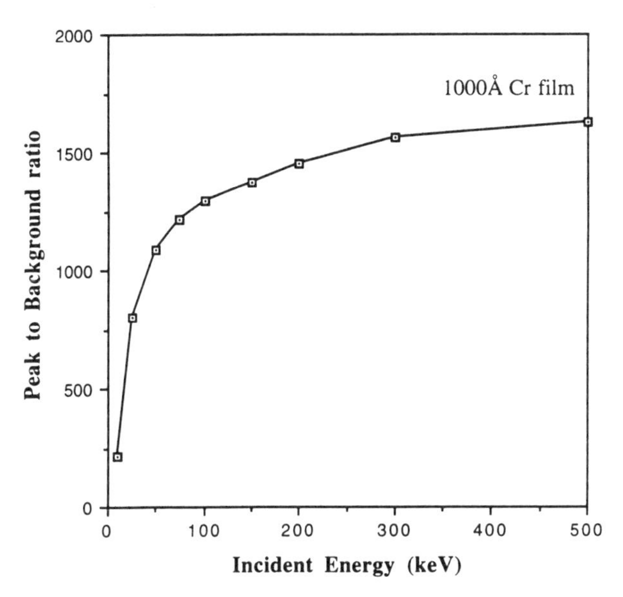

Figure 9.5. Predicted P/B ratio using Kramer's law model.

anisotropy. The most convenient approach is to employ the numerical evaluation and fit of Sommerfeld's *Bremsstrahlung* cross-section data (1931) made by Kirkpatrick and Wiedmann (1945). The procedure given below uses is a modification by Blackson (private communication) of the routine given by Statham (1976), with the relativistic corrections of Scheer and Zeitter (1955) and Zaluzec (1978). The original, rather cryptic notation of Kirkpatrick and Wiedmann has been retained so as to help in identifying the purpose of the various steps in the code.

```
Procedure continuum (energy:real;stepsize:real);
   {calculates a relativistically corrected continuum intensity for a
10-eV window at E_bkg using the Kirkpatrick and Wiedmann evaluation
of Sommerfeld's cross section. The result is in continuum counts per
unit energy width per electron. This version adapted from an original
of J. Blackson}

var    position:integer;
        joe,y1,y2,y3,h,j,m,a,b,c,d,iover:real;
        Ia,Qr,Qs,Ec,Eo:real;
  begin
      position:=round (50*z/thick);
        if energy>E_bkg then {the continuum window can be excited}
              Ec:=E_bkg*1000; {convert from keV to eV}
               Eo:=energy*1000; {ditto}
                beta:=1+(energy/511);
```

```
                beta:=1/(beta*beta);
                 joe:=Eo/(300*at_num*at_num); {K-W scaling factor}
     {now evaluate the terms in the K-W numerical fit}
                 y1:=exp(-26.9*joe);
                 y1:=0.22*(1-(0.39*y1));
                y2: 0.067 + (0.023/(joe + 0.75));
                 y3:= 0.00259 + (0.00776/(joe+0.116));

          h:=((=0.214*y1) + (1.21*y2) -y3)/((1.43*y1) -
           (2.43*y2)+y3);
           j:=((1+(2*h))*y2) - ((2*(1+h))*ye);
            m:=(1+h)*(y3+j);

           b:=exp(-0.0828*joe) - exp(84.9*joe);
            a:=exp(-0.223*joe) - exp(-57*joe);
             c:= 1.47*b - 0.507*a - 0.833;
               d:= 1.7*b - 1.09*a - 0.627;
                    iover:=Ec/Eo;
        {get the components of the continuum in X and Y directions}
                     Ix:=c*(iover-0.135);
                     Ia:=d*(iover-0.135)*(iover-0.135);
                      Ix:=(0.252+Ix)-Ia;
                      Ix:=(1/joe)*Ix*1.51E-27; {X-component}

                     Iy:=iover + h;
                      Iy:=(1/joe)*(-j+(m/Iy))*1.51E-27;
                      {Y-component}
   {correct for the line of sight of detector to anisotropic
Bremsstrahlung}
                     den:=1 - beta*cos(TOA);
                       den:=den*den*den*den;
                     Qz:=cos(TOA);
                      Qz:=Qz*Qz;
                     Qr:=1-Qz;
                       Qs:=Qz/den;
                     Qz:=(Ix*(Qr/den))+(Iy*(1+Qs));
                    {so the net contribution is}
                      bkg_yield:=stepsize*Qz;
                    end {of the if statement}
                    else
                        bkg_yield:=0;
              {add this contribution to the histogram}
               if ((position>=0)and(position<=50)) then {within
                 array bounds}
                 bkg_ro_z[position]:=bkg-ro_z[position] + bkg
                 _yield;
  end;
```

In order to properly account for the anisotropy of the continuum, the takeoff angle of the detector must be known. In standard microanalysis nomenclature, the takeoff angle (TOA) is the angle between the line joining the detector to the beam point and the surface of the specimen. In order to match the Kirkpatrick-Wiedmann notation, this angle must be referenced to the incident beam direction. That is,

```
{code to input and properly compute the takeoff angle}
         GoToXY(40,9);
           write('Detector take-off angle (deg)');
              readln(TOA);
           {for compatibility with W-K notation}
              TOA:=(TOA + 90)/57.4; {in radians}
```

Finally, as before, the continuum and characteristic contributions must be summed, over depth, and the missing constants must be reinserted:

```
   {sum over depth}
         for k:=0 to 50 do
             begin
              the_peak_is:=the_peak_is + phi_ro_z[k];
              the_bkg_is:=the_bkg_is + bkg_ro_z[k];
             end;
   {put in the missing constants for both terms}
  the_peak_is:=100*the_peak_is*6.02E23*6.51E-20*(density/at
_wht)*1E-8;
  the_bkg_is:=100*the_bkg_is*6.023E23*1E-8*(density/at
_wht)*10*4*3.14159;
```

Using conditions identical with those tried above and choosing a detector takeoff angle of 30°, this model predicts a Fiori number at 100 keV of about 2400, again falling by about 5% as the film thickness is varied over the range 250 to 5000 Å. This number is in excellent agreement with current experimental data. Figure 9.6 show how the peak to local background for a 1000-Å chromium film (Fiori number) varies with incident beam energy. The trend is the same that found with the previous model, a rapid rise with energy up to 100 keV and then a slow increase. Figure 9.7 show how the Fiori number at 100 keV is changed by the choice of detector takeoff angle. As TOA is varied from 0 (i.e., looking at glancing incidence across the specimen surface) to 60°, the P/B ratio increases by about 50% because of the forward peak of the continuum, confirming the usual preference for a high-takeoff-angle detector.

Further sophistication in both the continuum and characteristic cross sections could be used to get still better data (e.g., Gray et al., 1983), in particular at the highest energies, where relativistic effects are significant. However, for most typical

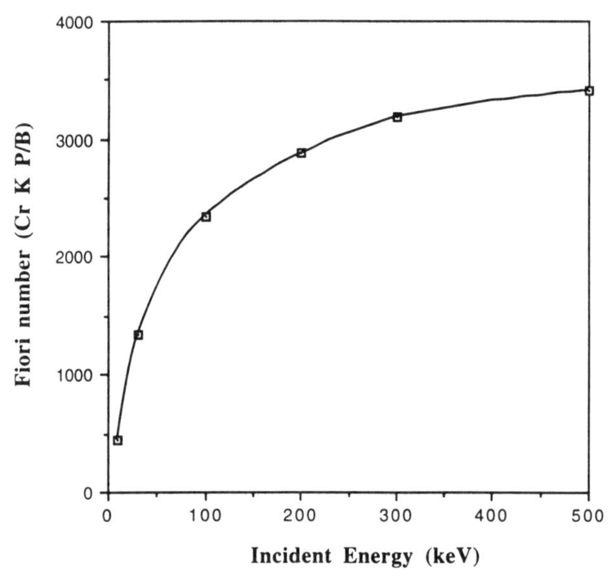

Figure 9.6. Computed P/B ratio vs. energy, using Wiedman-Kirkpatrick model.

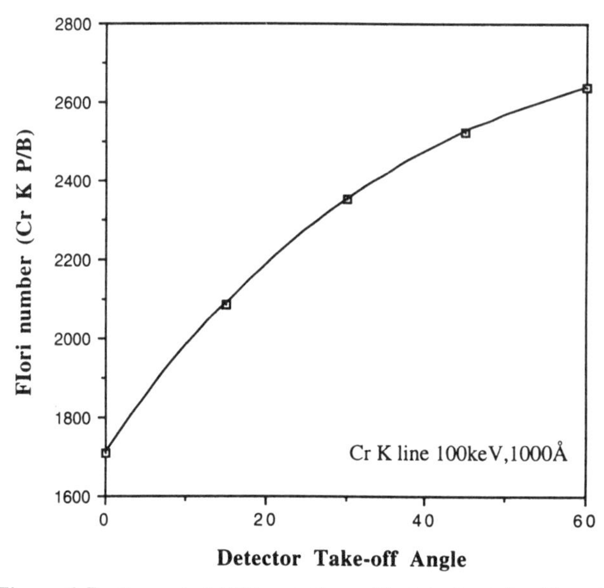

Figure 9.7. Computed P/B variation with detector takeoff angle.

operating conditions, the models given here are accurate enough to provide a good guide to what can be expected experimentally.

9.5 X-ray production in bulk samples

Extending the models discussed above to the case of a bulk specimen is straightforward, since exactly the same code can be used. As usual, however, the speed of a bulk simulation using a single scattering model may be unacceptably slow, so the plural scattering approach may be of more practical use. The foundation of all quantitative methods for bulk microanalysis is the depth dependence of the x-ray production, a function usually called $\phi(\rho z)$, following Castaing (1960), who defined $\phi(\rho z)$ as the x-ray generation in a slice $\delta(\rho z)$ of the specimen, of density ρ and at some depth z beneath the surface, normalized by the corresponding x-ray production from an identical freestanding slice. When presented in this way, the $\phi(\rho z)$ curves are found to be, to a good approximation, of a simple generic and analytical form that is sensibly dependent on the atomic number of the elements involved.

The computation of $\phi(\rho z)$ curves was one of the first tasks to which Monte Carlo methods were applied (Bishop, 1966), and much of the development of quantitation methods has relied on such computations (e.g., Curgenven and Duncumb, 1971; Duncumb, 1992). A calculation in the form specified by the original Castaing definition is possible, but the majority of computations have instead normalized the data by the value of $\phi(\rho z)$ at the surface of the specimen. There is than a constant ratio between the value that would be derived from Castaing's definition and the computed $\phi(\rho z)$, this ratio being the quantity $\phi(0)$, which is of the order of $1 + \eta$ where η is the backscattering coefficient of the bulk sample (Merlet, 1992). The program PhiRoZ computes and displays $\phi(\rho z)$ curve using the procedure `Generate_x_rays` discussed above and our usual plural scattering Monte Carlo model. Since the Bethe range of the incident beam is already divided into 50 steps in this model, these steps are used as the intervals for the $\phi(\rho z)$ histogram. As before, x-ray generation is assigned to the midpoint of the trajectory step. Once the energy of incident electrons has fallen below the critical excitation energy E_{crit} for the x-ray line of interest, the trajectory calculation can be aborted without error to save time. However, for illustrative purposes it is sometimes better to continue the computation but to identify portions of the trajectory that are above and below E_{crit}—for example, by color-coding the two portions of trajectory. In this way the relationship between the x-ray generation volume and the beam interaction volume can readily be seen. In the PHIROZ program on the disk, the approach used by Curgenven and Duncumb (1971) in their pioneer work is followed. A random weighting algorithm varying with the x-ray ionization cross section determines whether or not a dot will be plotted at the midpoint of each trajectory step for which the electron energy is above E_{crit}. This builds up a display in which dot density is proportional to ionization probability. Figure 9.8 shows a typical output from the program for the genera-

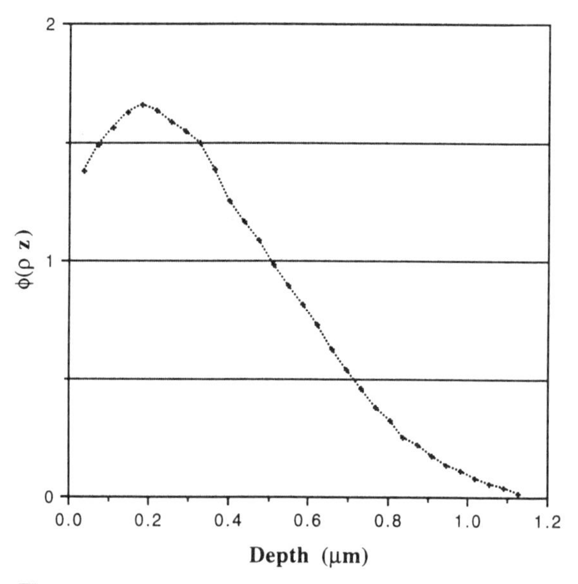

Figure 9.8. Computed $\phi(\rho\ z)$ curve for Fe K at 20 keV.

tion of Fe Kα x-rays at 20 keV. $\phi(\rho z)$ is plotted as a horizontal histogram on the same scale as the trajectories, so that the relationship between them can readily be seen. As the depth beneath the surface is increased, $\phi(\rho z)$ rises because the ionization cross section [Eq. (9.1)] continues to rise as E falls until $E \approx 2.5 \times E_{crit}$. Figure 9.9 shows some typical $\phi(\rho z)$ curves computed from this program for the excitation

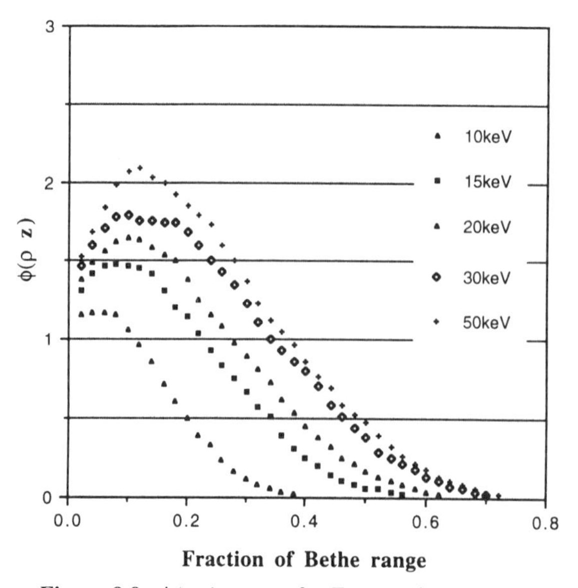

Figure 9.9. $\phi(\rho\ z)$ curves for Fe at various energies.

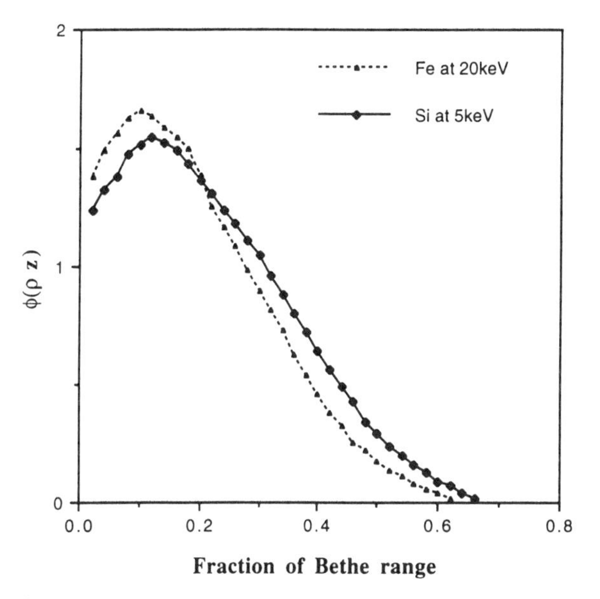

Figure 9.10. Computed curves for Fe at 20 keV, Si at 5 keV, giving same overvoltage.

of iron Kα at 10, 15, 20, 30, and 50 keV. The systematic change in the $\phi(\rho z)$ curves is evident. As the accelerating voltage is increased the peak magnitude of the profile rises, and moves deeper into the sample but the general shape of the profile remains the same. This is further illustrated in Fig. 9.10, which shows $\phi(\rho z)$ profiles for iron Kα at 20 keV and silicon K at 5 keV, which in both cases represents an overvoltage $U = E/E_{crit}$ of about 3. Note that the two curves are very similar in shape and size, even though the energy of the x-rays and the respective electron ranges differ by nearly an order of magnitude, demonstrating the value of the $\phi(\rho z)$ approach as a way of simplifying the representation of x-ray generation. A complete program for quantitative microanalysis can be constructed from this Monte Carlo approach (Duncumb, 1992).

Experimental $\phi(\rho z)$ profiles are generated by monitoring the x-ray production from a thin film of a "tracer" element embedded at different depths within the material of interest (Castaing, 1960), since only in this way can be the specific depth dependence of generation be separated from the total emission. However, because the x-ray line that is being studied in the tracer is not excited at the same energy as the desired line from the material, the form of the $\phi(\rho z)$ profile will not be identical with that expected from the element itself. By using the PHIROZ program, this problem can also be studied. Figure 9.11 shows two $\phi(\rho z)$ profiles, one computed for the Kα line of a Mn tracer in iron and the other for iron Kα in iron. The Mn profile peaks slightly sooner and slightly higher than the Fe curve, because the Mn line is of lower energy. A similar simulation for a tracer of higher energy than the

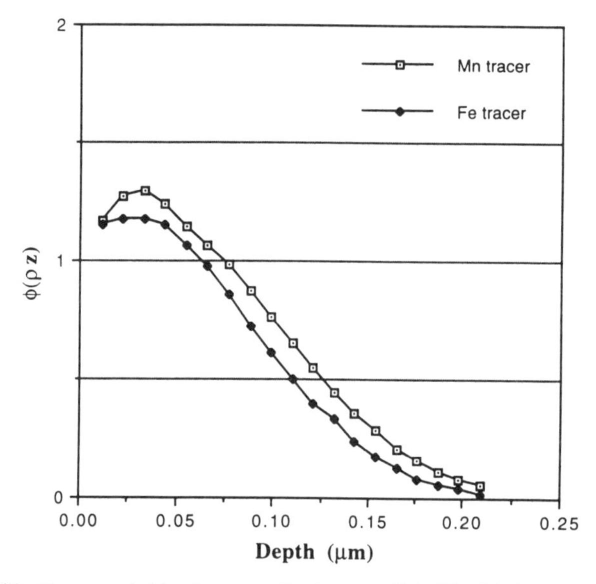

Figure 9.11. Computed $\phi(\rho\ z)$ curves for iron at 10 keV with Mn and Fe tracers.

iron would show the opposite situation. Thus, while the tracer method gives a good experimental approximation to the true $\phi(\rho z)$ profile it is not exact.

By substituting one of the `continuum` procedures given above, the corresponding depth variation of the *Bremsstrahlung* generation can also be studied. In this case the energy of the continuum window to be modeled, previously set to E_{crit}, can be varied to see how the depth dependence varies with energy. Figure 9.12

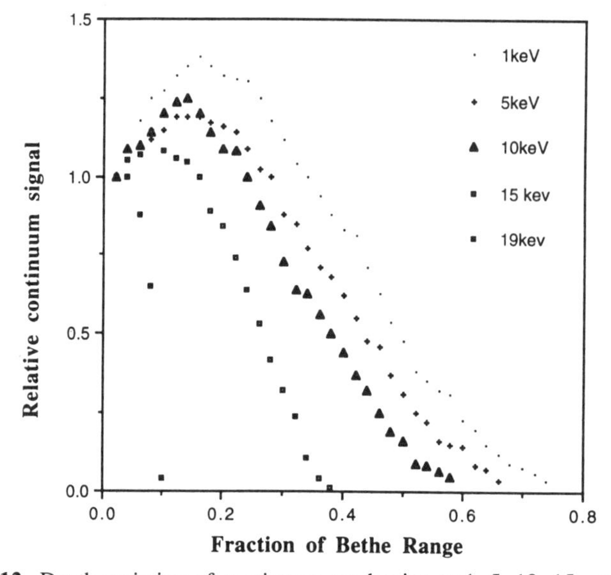

Figure 9.12. Depth variation of continuum production at 1, 5, 10, 15, and 19 keV.

shows a sequence of "$\phi(\rho z)$" plots obtained in this way for an iron sample at 20 keV and for continuum energies of 1, 5, 10, 15, and 19 keV obtained from the program BREMS on the disk, which implements this approach. All of the continuum windows are computed simultaneously from the same set of trajectories, since they are independent events. Note that the shape of the $\phi(\rho z)$ profile is different from that for characteristic radiations because of the different functional form of the cross section, and this is true even when the energy of the continuum window is the same as the energy of the characteristic line. Thus methods that rely on the ratio of the peak (i.e., characteristic line) to the background (i.e., continuum) at the same energy to correct for x-ray absorption in the sample (Statham and Pawley, 1978) will not generally produce a satisfactory result (Rez and Kanopka, 1984) unless corrected (Lu and Joy, 1992) to account for this discrepancy. Both Kramer's law and the Kirkpatrick and Wiedmann models for continuum production yield very similar results for bulk samples because the multiple scattering of the beam averages out the polarization of the continuum signal. Either can thus be used as desired.

The model can be further developed by computing the shape of the *Bremsstrahlung* as seen by an energy dispersive spectrometer (EDS) detector placed outside of the sample. This requires that the continuum be simulated for a number of energy windows between zero and the incident beam energy; here, 19 such regions equally spaced are used. Each of these results must then be corrected for x-ray absorption both in the specimen and in the EDS detector window. The absorption of x-rays is described by Beer's law, which gives the intensity $I(x)$ as some distance x from the point of production in the form

$$I(x) = I(0). \exp[-\mu\rho x] \tag{9.7}$$

where ρ is the density and μ is the mass-absorption coefficient. μ depends both on the material and on the energy of the x-ray passing through it. Tables of values are available (e.g., Goldstein et al., 1992), but since numerical fits through the data have been made by Heinrich, a simple routine can be used to compute the μ values that are needed with good accuracy.

```
Procedure abs_corr(photon_energy:real):real;
   {calculates the mass absorption coefficient using Heinrich's
method-the photon energy is in keV and the result is in cm^2/g}
var
   A,B,C,D,F,dum,dum2,dum3:real;

  begin
      if photon_energy<E_crit then {below K-absorption edge}
        begin
           A:=1.0;
            B:=-0.2544711;
```

```
            C:=4.769245;
          D:=-10.37878;
         F:=2.73
       end
     else {we are above the K-edge}
       begin
         A:=1.0;
          B:=-0.2322294;
           C:=4.070053;
           D:=-6.220746;
  F:=exp(-0.0045522*ln(at_num)*ln(at_num)-0.0068535*ln(at
_num)+1.070181);
       end;
   {compute the value using appropriate constants}
       dum:=B*ln)at_num)*ln(at_num)+C*ln(at_num)+D;
        dum2:=12.398/photon_energy;
         dum3:=power(dum2,F);
         abs_corr:=A*dum3*exp(dum);
  end;
```

If the TOA of the EDS detector is specified, then for an x-ray generated at depth z beneath the surface, the exit path length is z/sin (TOA). At the end of the simulation, therefore, the continuum "$\phi(\rho z)$" data is corrected in a stepwise fashion for each of the 19 continuum windows:

```
for i:=1 to 19 do
    begin {by getting energy of the i-th continuum window}
      photon_energy:=i*inc_energy/20;
    {get the mass absorption coefficient at this energy}
        factor:=abs_corr(photon_energy)/sin(TOA/57.4);

    for j:=0 to 50 do {sum over all layers down from surface}
      begin {compute exit path length × density × MAC}
        factor2:=factor*density*j*step*1E-4;
        factor3:=exp(-factor2); {Beer's law}
     {now compute what would be observed outside sample}
    bremss[i]:=bremss[i]+phi_ro_z[j,i]*factor3;
      end; {j-loop}
    end; {i-loop}
```

Finally, the effect of the detector on the spectrum must be accounted for. Over most of the energy range, the dominant effect is absorption in the beryllium window in front of the Si(Li) diode, so this can be found by using the same routine as before, but replacing the physical constants of the sample with those of the beryllium window:

```
Function det_factor(wind_z,wind_thick,wind_density,energy:real):real;
   {computes EDS detector efficiency given the window thickness,
   atomic number and density}
var
   factor,factor2,factor3:real;
    begin
      at_num:=wind_z; {for abs_corr routine}
        factor:=abs_corr(energy);
      {density*path length through window*MAC}
         factor2:=factor*wind_density*wind_thick*1E-4;
          factor3:=exp(-factor2); {Beer's law}
           det_factor:=factor3;
    end;
```

thus we have

```
   for i:=1 to 19 do {the 19 continuum windows}
    begin
      photon_energy:=i*inc_energy/20.0;
   {putting in constants for a Be window 6 μm thick, Z = 4, ρ = 1.2}
      det_effic[i]:=det_factor(4,6,1.2,photon_energy);
     {now correct data for this additional absorption}
         bremss[i]:=bremss[i]*det_effic[i];
    end;
```

The final step is then to interpolate data points linearly between the calculated values so as to simulate the whole spectrum. Figure 9.13 shows the result of a calculation for the *Bremsstrahlung* spectrum of carbon at 20 keV. The dots show the corresponding experimental data, scaled to match at the peak of the continuum. As can be seen, the agreement between the normalized and experimental spectra is excellent, even though the model is simple; although the ratio of absolute predicted and measured intensities is less good. The *Bremsstrahlung* spectrum is not often considered as being worthy of examination and is usually removed mathematically before any analysis of the characteristic peaks is made. However, with the aid of a model such as we have developed here, the behavior of the continuum can conveniently be studied, and we find that this component of the spectrum can provide much useful information, especially because the *Bremsstrahlung* covers a wide range of energies. First, the area beneath the continuum after correction for x-ray absorption varies linearly with the atomic number of the target, so providing a simple check on any qualitative analysis that is performed. Second, the shape of the continuum depends on the absorption experienced by the photons as they travel to the surface. Thus the shape depends on the general shape of the surface—i.e., its topography. This is important in analyzing particles, samples containing edges, or samples with a large degree of surface topography, since the form of the continuum

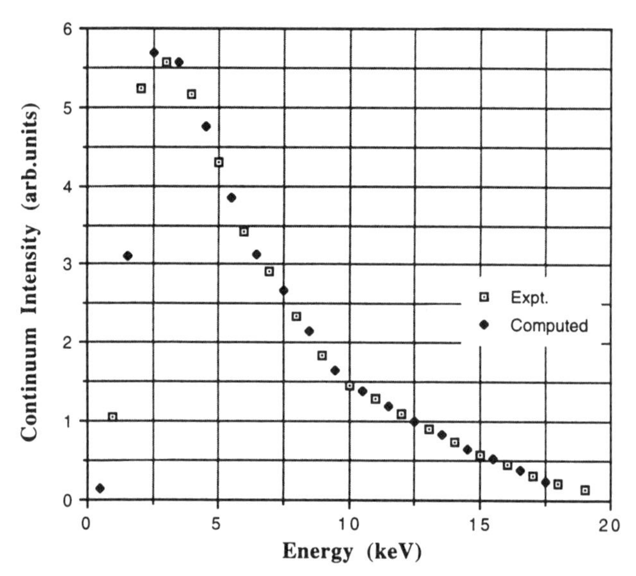

Figure 9.13. Comparison of experimental and computed *Bremsstrahlung* for copper at 20 keV, 8-μm Be window.

can be used both to understand what is happening in the sample and then to correct for this. For example, the position of the peak in the *Bremsstrahlung* depends on the length of the aborption path to the surface. If the effective radius of curvature of the surface around the beam entrance point is reduced from infinity (i.e., a flat surface) down to values of a few micrometers, then the peak shifts smoothly. Thus a measure of the surface conformation can be obtained from an observation of the continuum. Similarly, the presence of charging fields in the specimen will systematically distort the form of the *Bremsstrahlung* profile.

The techniques discussed above can readily be extended to deal with any other geometrical condition. For example, the important case of performing x-ray analysis of a thin film on a substrate can be modeled by using the procedures given in Chap. 6 to compute the $\phi(\rho z)$ profiles in the two materials and then accounting for the x-ray absorption of the substrate radiation in the film by using the routine given earlier in this chapter. In general, it is sufficient to employ a plural scattering routine to compute the trajectories, but in any case where a region of interest is less than about 5% of the electron range, it would be better to use the alternative single scattering model to ensure better accuracy.

10

WHAT NEXT IN MONTE CARLO SIMULATIONS?

10.1 Improving the Monte Carlo model

The stated aim of this book was to develop Monte Carlo simulations that were accurate enough to be useful predictive tools but simple enough to be accessible to nonspecialists. The previous chapters of the book have demonstrated that, within these limits, a great deal can be done. However, in meeting these goals, approximations and simplifications have had to be made. Specifically, we have made the assumptions that only elastic scattering, as described by the Rutherford cross section, need be considered and that energy loss can be treated as a continuous rather than as a discrete process. Such assumptions are not essential to the construction of a Monte Carlo simulation of electron scattering, and so the obvious and desirable move is to remove these approximations for those cases where they represent an unacceptable restriction.

An important first step is to use the more accurate description of elastic scattering provided by the Mott, or partial wave expansion, cross section. As noted in Chap. 3, the Rutherford cross section works well for the scattering of medium to high-energy electrons by nuclei of low atomic number. However under some conditions, particularly at low beam energies ($E < 20$ keV) and for high-atomic-number elements ($Z > 30$), where the scattering angles are large, the Rutherford formula is only an approximation. When the cross section is derived from the relativistic Dirac equation, taking into account spin-orbit coupling, then the "Mott" scattering cross section which results can differ significantly from the Rutherford values. Such discrepancies become important, and experimentally detectable, in cases where each electron is scattered only a few times and where the angular distributions of transmitted or scattered electrons of various energies are required to be known accurately.

The reason why the Mott cross section, despite its superiority, has not been more widely used for Monte Carlo simulations is that it is not an analytical function—i.e., one that can be evaluated from an equation—but an array of differential cross-section values representing the scattering probability for electrons of a particular energy in a specified direction. Typically (Czyzewski et al., 1990) the Mott cross section must be evaluated at energy steps of 100 eV or so from 20 eV to 20 keV and for angular increments of 1 or 2° from 0 to 180°, giving a total of some

2500 sets of numbers for each element. Storing the data for the entire periodic table therefore requires several megabytes of storage space on disk. Angular scattering probabilities and total cross sections must then be found by numerical integrations of these data arrays. While these operations are not intrinsically difficult, they are messy and tend to make the programs that employ them look rather complex compared to the normal run of personal computer software.

One approach that has been successfully implemented to obtain the benefits of the Mott cross section without the complexity is to reduce it to an analytical form by a process of parametrization. Browning (1992) showed that the total elastic Mott cross section σ_M could be written as

$$\sigma_M = 4.7 \times 10^{-18} \cdot \frac{(Z^{1.33} + 0.032\,Z^2)}{(E + 0.0155Z^{1.33} \cdot E^{0.5})} \cdot \frac{1}{(1 - 0.02Z^{0.5} \cdot e^{-u^2})} \tag{10.1}$$

where $u = \log_{10}(8.E.Z^{-1.33})$

Equation (10.1) is a good description of the overall trends of the cross section, although it cannot follow the periodicity of the Mott cross sections, which are directly due to periodic variations of the size of the atoms. Such effects have little macroscopic effect on the scattering of electrons by a solid, however. This parametrized cross section can be substituted directly for the usual Rutherford cross section in the single scattering program. Since the Mott cross section is typically smaller than the corresponding Rutherford value, the elastic mean free paths that are computed will be larger and the number of scattering events that must be computed will therefore generally be smaller, leading to a useful improvement in program speed.

It is also necessary for the computed angular scattering distribution to be modified to match the results of the Mott cross section. As shown in Fig. 10.1, the cumulative angular scattering distribution for the Mott model is rather more square in form than the Rutherford distribution. This behavior can be approximated by modifying the form of the Rutherford equation, e.g., Equation (3.10):

$$\cos \phi = 1 - \frac{2\alpha\,\mathrm{RND}}{(1 + \alpha - \mathrm{RND})} \tag{3.10}$$

to be

$$\cos \phi = 1 - \frac{2\alpha.\mathrm{RND}^2}{(1 + \alpha - \mathrm{RND})} \tag{10.2}$$

which is a good match to the exact numerical data over much of the periodic table.

Parametrized models are a useful step up from the simplest Rutherford models and should, with further work, be capable of giving good-quality data. It must, however, be realized that such an approach is only an approximation and that under

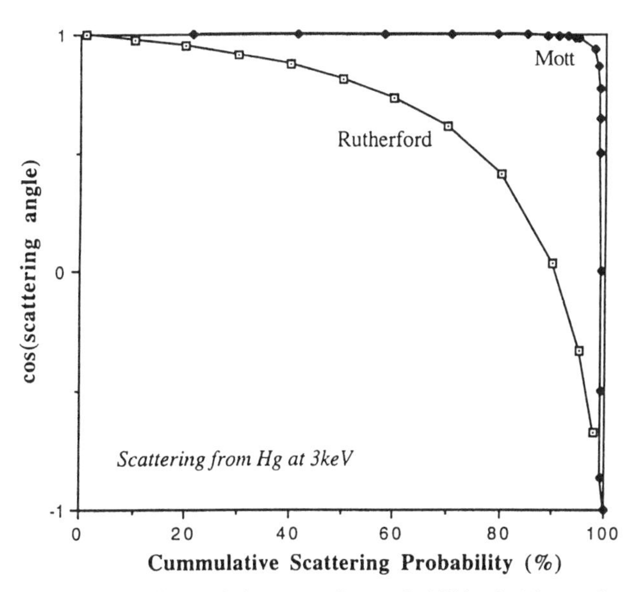

Figure 10.1. Comparison of cumulative scattering probabilities in Mott and screened Rutherford models.

extreme conditions—such as very low voltages (<1 keV) and at high scattering angles—the approximation will not be accurate. When such conditions must be explored, it is therefore necessary to go to a full Mott model, using numerically tabulated differential cross-section data. Many such programs have been described (e.g., Reimer and Stelter, 1987; Kotera et al., 1990; Czyzewski and Joy, 1989), and tabulated sets of Mott cross sections have been published by several groups (Reimer and Lodding, 1984; Czyzewski et al., 1990). Such programs have been demonstrated to give excellent results for both macroscopic events, such as the variation of backscattering with incident energy, and microscopic data, such as the angular and energy distributions of scattered and transmitted electrons (Reimer and Krefting, 1976). With the continued increase in the memory capacity of personal computers, Mott models will certainly become more commonly applied because of the undoubted improvements that they can provide in some circumstances. However, care must be taken that all aspects of the Monte Carlo program that is developed are equally advanced. Some published programs that have gone to great lengths to use the Mott cross section have compromised the quality of their results by using poor electron stopping-power formulations or by terminating the simulation at relatively high energies and relying on diffusion models to account for the rest of the trajectory.

A second important area of research activity has been to construct Monte Carlo simulations in which all types of scattering, elastic and inelastic, are considered and in which energy losses are treated as discrete rather than continuous events (e.g.,

Shimizu and Ichimura, 1981). Such models considerably extend the boundaries within which Monte Carlo modeling can be considered useful, although at the expense of increased complexity and computation time. The procedure followed is a generalization of that discussed in Chap. 7 for the modeling of fast secondary electrons. In addition to the usual elastic mean free path λ_E, a total inelastic mean free path λ_i is defined, where

$$\frac{1}{\lambda_i} = \frac{1}{\lambda_1} + \frac{1}{\lambda_2} + \frac{1}{\lambda_3} \quad \cdots\cdots\cdot \tag{10.3}$$

and the λ_j (j = 1, 2, etc.) are the mean free paths for each of the inelastic events considered (Fitting and Reinhart, 1985). The total mean free path is λ_T, where

$$\frac{1}{\lambda_T} = \frac{1}{\lambda_E} + \frac{1}{\lambda_i} \tag{10.4}$$

and for each scattering event a random number RND is compared with the ratio λ_T/λ_E to determine if the event is elastic or inelastic; then a second random number is drawn to decide what type of inelastic event is occurring. The usual Bethe law expression giving the electron stopping power is replaced by the sum of the energy lost in the inelastic events $\Sigma\ (\Delta E_i)$ encountered. Since the mixture of events along each trajectory will be different, the energy losses will correspondingly vary, leading to range straggling—the phenomenon of a statistical variation in electron range for a fixed incident energy—just as is observed experimentally. A program of this type is needed if information about the energy and angular distribution of electrons is sought, for example, to model Auger or energy loss spectra (Shimizu and Ichimura, 1981).

10.2 Faster Monte Carlo modeling

For many people the problem with Monte Carlo simulations has been and remains that it is a sequential procedure. Once the model has been constructed, it must be run a large number of times one after another in order to achieve the desired statistical accuracy before the data can be extracted. Although the speed of all computers continues to increase, the necessity of running 5000 or more trajectories per data point can still represent a major investment in time when perhaps 50 or more points must be modeled or when a large number of different sets of experimental parameters must be compared. Simply increasing computer speeds still further is a brute-force approach that can offer only a limited degree of improvement, especially since the very fastest machines tend to be multiuser computers that allocate to each program only some fraction of the available processor time. For example, when the time required for data input and output on a typical time-sharing system is included, a Cray supercomputer is less than twice as fast in running the same code as a 33-Mhz 486-type MS-DOS computer equipped with a math coprocessor chip. One solution to this dilemma is to run more than one trajectory at a time—that is, to use parallel computing techniques. Monte Carlo methods are well suited to this approach because the code is relatively compact and each trajectory is independent of every other. It is therefore straightforward to achieve a large gain in throughput simply by running one copy of the program on each of the processors in a parallel machine. Several examples of this have been reported (e.g., Michaels et al., 1993) on machines employing up to 128 processors. Impressive results, combining both good statistics and many data points, have been demonstrated for situations such as the modeling of x-ray generation across a complex interface and for the production of SE image simulations. The actual improvement in speed, while dramatic, is usually somewhat less than the number of CPUs used in parallel, because the number of computations required for each trajectory is not a constant. One processor may therefore finish all of its allotted simulations while another is still working, and unless the computer can reorganize its workload to reoccupy all of the available processors, the speed advantage gradually decreases as more and more CPUs finish their tasks and fall idle. Nevertheless, with computers that have more and more parallel architecture becoming available, this will be an important avenue of progress in this field.

10.3 Alternatives to sequential Monte Carlo modeling

A typical use of the kind of Monte Carlo models developed in this book is to investigate the effect of varying one or more parameters in some experimental setup; for example, seeing how changing the position of a detector affects the form of a backscattered electron signal profile or how altering the morphology of a surface might vary the x-ray yield. In such cases, even though almost all of the conditions remain the same, it is still necessary to rerun the entire simulation over again in order to get the new data, which is a wasteful and time-consuming procedure. An interesting alternative procedure has been described (Desai and Reimer, 1990; Wang and Joy, 1991; Czyzewski and Joy, 1992), which is a mixture of a conventional Monte Carlo model and an alternative physical representation called a "diffusion matrix."

Consider the problem of computing the backscattering coefficient of a flat, semi-infinite target. Suppose this specimen to be divided up into a larger number of cubes, each of which has the property of emitting electrons at a certain rate and with some known angular distribution. The measured yield of backscattered electrons will result from the emissions coming from those cubes with one face on the top surface of the specimen. If the topography, for example, is now changed—perhaps by placing a groove across the surface—then some additional cubes will have faces at the surface and the number of emitted electrons will change.

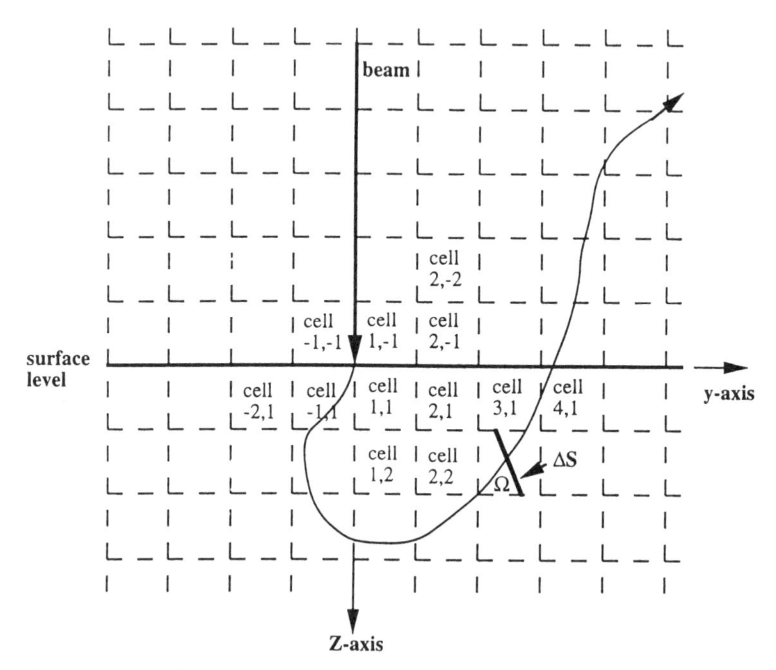

Figure 10.2. Layout of matrix. The Monte Carlo simulation finds how many BSE with a specified exit angle pass through a strip of width ΔS at angle Ω to the horizontal in each cell.

The strength of each elemental cuboid emitter of electrons is found by running a Monte Carlo simulation for the chosen experimental conditions. However, unlike a conventional simulation, the incident beam is considered to be emerging from a point inside the specimen (Fig. 10.2) rather than impinging at a surface. Every trajectory is followed until the electron completely dissipates its energy, because there is now no surface from which to be backscattered. The interaction volume is again divided into cubes, both above and below the beam point, and the Monte Carlo is configured so that for the cube X, Y, Z relative to the beam at coordinates (0, 0, 0), the number and energy of electrons crossing the faces of the cube at various angles relative to the axes is determined. The resultant multidimensional array then completely specifies the matrix of cubes. (The name *diffusion matrix* refers to an analogous model in which a point diffusion source—such as a source of heat—is replaced by an array of emitters that gives the same resultant properties). Typically the interaction volume is divided into $64 \times 64 \times 32$ cubes, the angular variation into 20 steps, and the energy into 10 steps. As usual, the Monte Carlo simulation must be continued for a sufficient number of trajectories so that each element of the array has adequately good statistics. The array can be further extended by including components representing the corresponding fluxes of SE or x-rays, and some initial research has demonstrated the feasibility of considering even multicomponent specimens.

Once this has been done, the yield of electrons for any arbitrary surface geometry or for any specified detector position relative to the specimen can be found by simply summing the relevant contributions from all the surface cubes. This makes it possible to almost instantaneously compute, for example, the effect of tilt on the backscattered yield or the SE profile from a surface of given geometry. In an implementation on a Macintosh IIci machine, the time to recompute a 256-pixel surface profile for a specified surface geometry was less than 1 sec, so this technique is well suited for studies in which the computed profile is to be interactively matched to an experimental profile. For many applications, this method has a great deal of promise, since it offers a good balance between speed and accuracy even on small computers. The diffusion matrix method is, of course, an approximation rather than an exact solution, since it assumes that each cubical element has properties that are fixed and independent of its surroundings and position relative to a surface. The method also cannot take account of effects such as the recollection, or scattering, of emitted electrons from the sample itself, which may be significant in congested geometries. However, it is a valuable extension of the standard Monte Carlo approach and can produce a lot of additional information for little extra computation.

10.4 Conclusions

Simulations will become an increasingly important and effective tool for problem solving in electron microscopy and microanalysis. With further improvements in both computers and the software that is run on them, it is not too fanciful to envisage a "virtual reality" approach to electron microscopy in which fast, interactive, simulations can be used to test alternative experimental methods, to investigate the effect of different experimental parameters, to optimize conditions, and even produce "images" and "spectra" without the need to use expensive time on a real microscope. Even at a less exalted level, the simulations discussed in this book provide a convenient, flexible, and accurate way of interpreting data and images, a useful resource for planning experiments, and a satisfying teaching tool.

REFERENCES

Akamatsu B., Henoc J., and Henoc P. (1981), *J. Appl. Phys.*, 52:7245
Antolak A. J. and Williamson W. (1985), *J. Appl. Phys.*, 58:525–31
Archard G. D. (1961), *J. Appl. Phys.*, 32:1505–9
Austin L. and Starke H. (1902), *Ann. Phys. (Leipzig)*, 9:271
Baroody E. M. (1950), *Phys. Rev.*, 78:780
Berger M. J. (1963), in *Methods in Computational Physics*, vol. 1 (Eds. B. Adler, S. Fernbach, and M. Rotenberg), (Academic Press: New York), p. 135
Berger M. J. and Seltzer S. M. (1964), *Studies in Penetration of Charged Particles in Matter*, Nuclear Science Series Report #39, NAS-NRC Publication 1133 (Natl. Acad. Sci: Washington, D.C.), p. 205
Bethe H. A. (1930), *Ann. Phys.*, 5:325
Bethe H. A. (1941), *Phys. Rev.*, 59:940
Bishop, H. E. (1966), "Measurements and Calculations of Electron Backscattering." Ph.D. thesis, University of Cambridge
Bishop H. E. (1965), *Proc. Phys. Soc.*, 85:855
Bishop H. E. (1974), *J. Phys. D: Appl. Phys.*, 7:2009.
Bishop H. E. (1976), in *Use of Monte Carlo Calculations in Electron Probe Microanalysis and Scanning Electron Microscopy* (Eds. K.F.J. Heinrich, D. E. Newbury, and H. Yakowitz (U.S. Dept. of Commerce/National Bureau of Standards), NBS Special Publication #460, p. 5
Bradley G. F. and Joy D. C. (1991), Proc. 49th Ann. Meeting EMSA (San Jose), (Ed. G. W. Bailey), (San Francisco Press: San Francisco), p. 534
Bresse J. F. (1972), Proc. 5th Ann. SEM Symposium (Ed. O. Johari and I. Corvin), (Illinois Institute of Technology Research Institute: Chicago), 105
Bronstein I. M. and Fraiman B. S. (1961), *Sov. Phys. Solid State*, 3:1188
Browning R. (1992), *Appl. Phys. Lett.*, 58:2845
Bruining H. (1954), *Physics and Applications of Secondary Electron Emission*, (Pergamon Press: London)
Cailler M. and Ganachaud J. P. (1972), *J. Physique*, 33:903
Campbell-Swinton A. A. (1899), *Proc. Roy. Soc.*, 64:377
Casnati E., Tartari A., and Baraldi C. (1982), *J. Phys. B*, 15:155
Castaing R. (1960), "Electron probe microanalysis," in *Advances in Electronics and Electron Physics*, 13:317
Catto C.J.D. and Smith K.C.A. (1973), *J. Microsc.*, 98:417
Chung M. and Everhart T. E. (1977), *Phys. Rev. B*, 15:4699
Compton A. H. and Allison S. K. (1935), *X-rays in Theory and Equipment*, (Van Nostrand: New York)
Curgenven L. and Duncumb P. (1971), Tube Investments Ltd., Research Report #303

Czyzewski Z. and Joy D. C. (1989), *J. Microsc.*, 156:285
Czyzewski Z. and Joy D. C. (1991), *Scanning,* 13:227
Czyzewski Z. and Joy D. C. (1992), in Proc. 50th Ann. EMSA Meeting, Boston (Ed. G. W. Bailey, J. Bentley, and J. A. Small), (San Francisco Press: San Francisco), 954
Czyzewski Z., MacCallum D. O., Romig A. D., and Joy D. C. (1990), *J. Appl. Phys.*, 68:3066
Darlington E. H. (1975), *J. Phys. D,* 8:85
Dekker A. J. (1958), *Solid State Phys.*, 6:251
Desai V. and Reimer L. (1990), *Scanning,* 12:1
Devooght J., Dubus A., and Dehaes J. C. (1987), *Phys. Rev. B,* 36:5093
Donolato C. (1978), *Optik,* 52:19
Donolato C. (1985), *Appl. Phys. Lett.*, 46:270
Donolato C. (1988), Proc. Eur. Congr. on EM, *Inst. Phys. Conf. Ser.*, 93, 2:53
Drescher H., Reimer L., and Seidel H. (1970), *Z. Angew. Phys.*, 29:331–36
Duncumb P. (1977), personal communication cited on page 12 of Love et al. (1977)
Duncumb P. (1992), Proc. 50th Ann. Meeting EMSA (Boston), (Ed. G. W. Bailey, J. Bentley, and J. A. Small), vol. 2, p. 1674
Duntemann J. (1987), *Complete Turbo Pascal,* (Scott Foresman and Co.: San Francisco)
Dwyer V. M. and Matthew J.A.D. (1985), *Surface Sci.*, 152:884
Egerton R. F. (1986), *Electron Energy Loss Spectrometry in the Electron Microscope* (Plenum Press: New York)
Evans R. D. (1955), *The Atomic Nucleus* (McGraw Hill: New York), p. 576
Everhart T. E. (1960), *J. Appl. Phys.*, 31:1483–90
Fiori C. E., Swyt C. R., and Ellis J. R. (1982), in *Microbeam Analysis 1982* (Ed. K.F.J. Heinrich), (San Francisco Press: San Francisco), p. 57
Fitting H. J., Glaefeke H., and Wild W. (1977), *Phys. Stat. Sol. (a),* 32:185
Fitting H. J. and Reinhardt J. (1985), *Phys. Stat. Sol. (a),* 88:245
Gasper J. K. and Greer R. T. (1974), in *Scanning Electron Microscopy 1974* (ed. O. Johari), p. 244
Goldstein J. I., Costley, J. L., Lorimer G. W., and Reed S.J.B. (1977), Proc. 10th Ann. SEM Symposium (Ed. O. Johari), (SEM Inc.: Chicago), vol. 1, p. 315
Goldstein J. I., Newbury D. E., Echlin P., et al. (1992), *Scanning Electron Microscopy and X-ray Microanalysis* (New York: Plenum)
Gray C. C., Chapman J. N., Nicholson W.A.P., et al. (1983), *X-ray Spectrometry,* 12:163
Green M. (1963), *Proc. Phys. Soc.,* 82:204
Gryzinski M. (1965), *Phys. Rev. A,* 138:336
Hammersley J. M. (1964), *Monte Carlo Methods* (New York: Wiley)
Hasegawa S., Iida Y., and Hidaka T. (1987), *J. Vac. Sci. technol.*, B5:142
Hayward E. and Hubbell J. (1954), *Phys. Rev.*, 93:955
Hebbard D. F. and Wilson P. R. (1955), *Austral. J. Phys.*, 8:90
Heinrich K.F.J. (1981), *Electron Beam X-ray Microanalysis* (Van Nostrand: New York)
Heinrich K.F.J., Newbury D. E., Yakowitz H., eds., (1976), *Use of Monte Carlo Calculations in Electron Probe Microanalysis and Scanning Electron Microscopy* (U.S. Dept. of Commerce/National Bureau of Standards), NBS Special Publication #460, p. 5
Herrmann R. and Reimer L. (1984), *Scanning,* 6:20
Hohn F. J., Kindt M., Niedrig H., and Stuth B. (1976), *Optik,* 46:491
Holt D. B. and Joy D. C. (1989), "SEM Microcharacterization of Semiconductors" (Academic Press: London)
Hunger H.-J. and Küchler L. (1979), *Phys. Stat. Sol. (a),* 56:K45
ICRU (1983), *Stopping Powers of Electrons and Positrons,* Report #37 to International Committee on Radiation Units (ICRU: Bethesda, Md.)
Jakubowicz A. (1982), *Solid State Electronics,* 25:651

Jonker J.L.H. (1952), *Philips Res. Rep.*, 7:1
Joy D. C. (1983), *Microelectr. Eng.*, 1:103
Joy D. C. (1984), *J. Microsc.*, 136:241
Joy D. C. (1986), *J. Microsc.*, 143:233
Joy D. C. (1987a), *J. Microsc.*, 147:55
Joy D. C. (1987b), in *Microbeam Analysis—1987* (ed. R. H. Geiss), (San Francisco Press: San Francisco), p. 117
Joy D. C. (1989a), in *Computer Simulation of EM Diffraction and Images* (ed. W. Krakow and M. O'Keefe), (Minerals, Metals and Materials Society: Warrendale, Pa.), 209
Joy D. C. (1989b), *Hitachi Instrument News*, 16:3
Joy D. C. (1991), *Ultramicroscopy*, 37:216
Joy D. C. (1993), *Microbeam Analysis*, 2:S17. Copies of this database of experimental data on electron-solid interactions are available on request from the author.
Joy D. C. and Luo S. (1989), *Scanning*, 11:176
Joy D. C., Maher D. M., and Farrow R. C. (1985), Mat. Res. Soc. Symposia (Materials Research Society: Warrendale, Pa.), vol. 69, p. 171
Kahn H. (1950), *Nucleonics*, May 27
Kanaya K. and Ono S. (1984), in *Electron Beam Interaction with Solids* (Eds. D. Kyser, D. E. Newbury, H. Niedrig, and R. Shimizu), (SEM Inc.: AMF O'Hare, Ill.), p. 69
Kanter H. (1961), *Phys. Rev.*, 121:681
Kirkpatrick P. and Wiedmann L. (1945), *Phys. Rev.*, 67:321
Koshikawa T. and Shimizu R. (1974), *J. Phys. D: Appl. Phys.*, 7:1303
Kotera M., Kishida T., and Suga H. (1990), *Scanning Microsc.* Suppl., 4:111
Kyser D. (1979), in *Introduction to Analytical Electron Microscopy* (Ed. J. J. Hren, J. I. Goldstein, and D. C. Joy), (Plenum Press: New York), p. 199
Lander J. J., Schreiber H., Buck T. M., and Matthews J. R. (1963), *Appl. Phys. Lett.*, 3:206
Leamy H. J. (1982), *J. Appl. Phys.*, 53:R51
Leamy H. J., Kimerling, L. C., and Ferris S. D. (1976), Proc. 9th SEM Symposium (Ed. O. Johari), (SEM Inc.: Chicago), p. 529
Leiss J. E., Penner S., and Robinson C. S. (1957), *Phys. Rev.*, 107:1544
Love G., Cox M.G.C., and Scott V. D. (1977), *J. Phys. D: Appl. Phys.*, 10:7
Lu Y. and Joy D. C. (1992), Proc. 50th. Ann. Meeting EMSA (Boston), (Ed. G. W. Bailey), (San Francisco Press: San Francisco), vol. 2, p. 1678
Luo S. and Joy D. C. (1990), *Scanning Microscopy Suppl.*, 4:125
Luo S., Zhang Y., and Wu Z. (1987), *J. Microsc.*, 148:289
McCracken D. D. (1955), *Sci. Am.*, 192:90
Merlet C. (1992), *X-ray Spectrometry*, 21:229
Metropolis N. and Ulam S. (1949), *J. Am. Statis. Assoc.*, 44:335
Meyer H. A., ed. (1956), *Symposium on Monte Carlo Methods* (Wiley: New York)
Michael J. R., Plimpton S. J., and Romig A. D. (1993). *Ultramicroscopy*, 51:160
Mil'stein S., Joy D. C., Ferris S. D., and Kimerling L. C. (1984), *Phys. Stat. Sol. (a)*, 84:363
Moller C. (1931), *Z. Phys.*, 70:786
Mott N. F. (1930), *Proc. Roy. Soc. London A*, 126:259
Müller R. H. (1954), *Phys. Rev.*, 93:891
Murata K., Kyser D. F., and Ting C. H. (1981), *J. Appl. Phys.*, 52:4396
Murata K., Matsukawa T., and Shimizu R. (1971), *Jpn. J. Appl. Phys.*, 10:678
Myklebust R. L., Newbury D. E., and Yakowitz H. (1976), in *Use of Monte Carlo Calculations in Electron Probe Microanalysis and Scanning Electron Microscopy* (Ed. K.F.J. Heinrich, D. E. Newbury, and H. Yakowitz), (U.S. Dept. of Commerce/National Bureau of Standards), NBS Special Publication #460, p. 105
Neubert G. and Rogaschewski S. (1980), *Phys. Stat. Sol. (a)*, 59:35

Newbury D. E. and Myklebust R. L. (1981), in *Analytical Electron Microscopy 1981* (Ed. R. H. Geiss), (San Francisco Press: San Francisco), p. 91
Neidrig H. (1982), *J. Appl. Phys.,* 53:R15
Possin G. E. and Kirkpatrick C. G. (1979), Proc. 12th Ann. SEM Symposium (Ed. O. Johari), (SEM Inc: Chicago), 1:245
Possin G. E. and Norton J. F. (1975), Proc. 9th Ann. SEM Symposium (Ed. O. Johari), (IITRI: Chicago), p. 457
Postek M. T. and Joy D. C. (1986), *Solic State Technol.,* 12:77
Powell C. J. (1976), *Rev. Mod. Phys.,* 48:33
Powell C. J. (1984), in *Scanning Electron Microscopy 1984* (Ed. O. Johari), (SEM Inc.: Chicago), vol. 4, p. 1649
Press, W. H., Flannery B. P., Teukolsky S. A., and Vetterling W. T. (1986), *Numerical Recipes* (Cambridge University Press: New York).
Rao-Sahib T. S. and Wittry D. B. (1974), *J. Appl. Phys.,* 45:5060
Reimer L. (1993), "Image formation in Low Voltage SEM" (SPJE Optical Engineering Press: Bellingham WA)
Reimer L. and Krefting E. R. (1976), in *Use of Monte Carlo Calculations in Electron Probe Microanalysis and Scanning Electron Microscopy* (Ed. K.F.J. Heinrich, D. E. Newbury, and H. Yakowitz), (U.S. Dept. of Commerce/National Bureau of Standards), NBS Special Publication #460, p. 45
Reimer L. and Lodding B. (1984), *Scanning,* 6:128
Reimer L. and Stelter D. (1987), *Scanning,* 8:265
Reimer L. and Tolkamp C. (1980), *Scanning,* 3:35
Rez P. and Kanopka J. (1984), *X-ray Spectrom.,* 13:33
Salow H. (1940), *Phys. Z.,* 41:434
Samoto N. and Shimiziu R. (1983), *J. Appl. Phys.,* 54:3855
Schou J. (1980), *Phys. Rev. B.,* 22:2141
Schou J. (1988), *Scanning Microscopy,* 2:607
Scheer M. and Zeitter E. (1955), *Z. Phys.,* 140:642
Seah M. P. and Dench W. A. (1979), *Surf. Interface Anal.,* 1:2
Seiler H. (1984), *J. Appl. Phys.,* 54:R1
Sercel P. C., Lebens J. A., and Vahala K. J. (1989), *Rev. Sci. Instrum.,* 60:3775
Shreider Yu A. (1966), *The Monte Carlo Method* (Pergamon Press: New York)
Shimizu R. (1974), *J. Appl. Phys.,* 45:2107
Shimizu R. and Ichimura S. (1981), *Quantitative Analysis by Auger Electron Spectroscopy* (Toyota Foundation: Tokyo, Japan)
Shimizu R., Murata, K., and Shinoda G. (1966), in *X-ray Optics and Microanalysis* (Eds. R. Castaing, P. Descamps, and J. Philbert), (Paris: Hermann), p. 127
Sidei T., Higasimura T., Kinosita K. (1957), *Mem. Fac. Engng. Kyoto Univ.,* 19:220
Sommerfeld A. (1931), *Ann. Physik.,* 11:257
Starke H. (1898), *Ann. Phys.,* 66:49
Statham P. J. (1976), *X-ray Spectrometry,* 5:154
Statham P. J. and Pawley J. B. (1978), in *Scanning Electron Microscopy—1978* (Ed. O. Johari and I. Corvin), vol. 1, p. 469
Suga H., Fujiwara T., Kanai N., and Kotera M. (1990), in Proc. XIIth Int. Congr. on EM (Eds. L. D. Peachey and D. B. Williams), (San Francisco Press: San Francisco), vol. 1, p. 410
Sugiyama N., Ikeda S., and Uchikava Y. (1986), *J. Electr. Microsc.,* 35:9
Tung C. J., Ashley J. C., and Ritchie R. H. (1979), *Surface Science,* 81:427
Wang X. and Joy D. C. (1991), *J. Vac. Sci. Techn.,* B9:3753
Williams D. B. and Steel E. B. (1987), in *analytical Electron Microscopy—1987* (Ed. D. C. Joy), (San Francisco Press: San Francisco), p. 228

Wilson P. R. (1952), *Phys. Rev.*, 86:261
Wittry D. B. and Kyser D. F. (1964), *J. Appl. Phys.*, 35:2439
Wittry D. B. and Kyser D. F. (1965), *J. Appl. Phys.*, 36:1387
Wolff P. A. (1954), *Phys. Rev.*, 95:56
Zaluzec N. (1978), Ph.D. thesis, University of Illinois

INDEX